The Cambridge Manuals of Science and
Literature

PLANT-LIFE ON LAND

CONSIDERED IN SOME OF ITS BIOLOGICAL ASPECTS

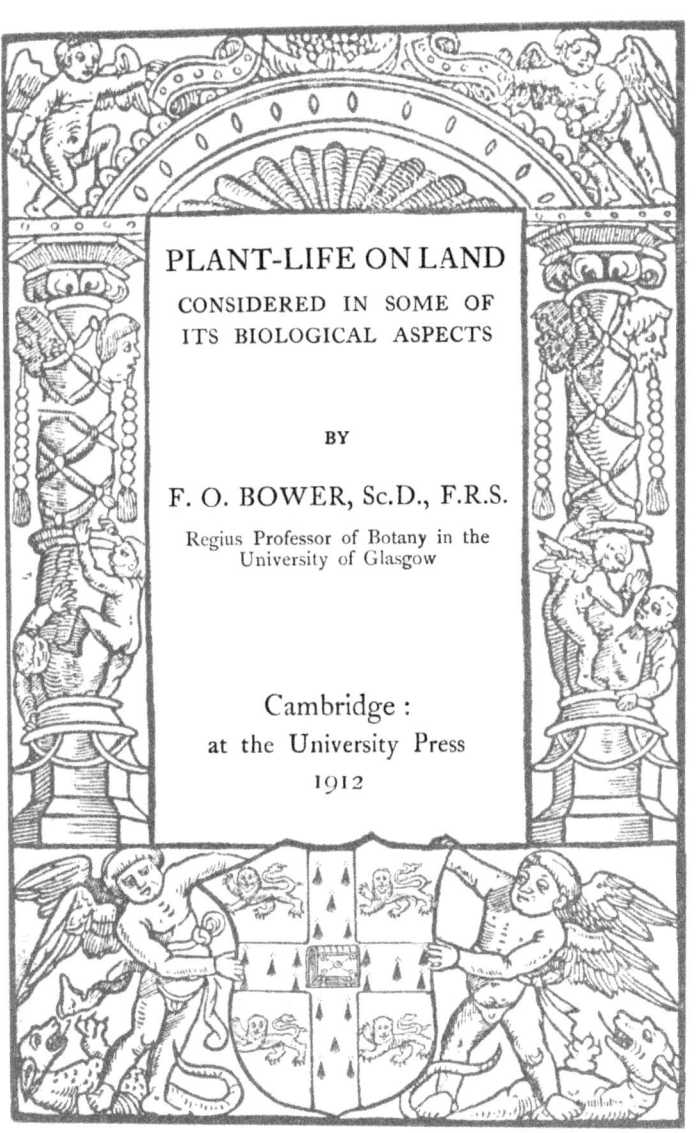

PLANT-LIFE ON LAND

CONSIDERED IN SOME OF
ITS BIOLOGICAL ASPECTS

BY

F. O. BOWER, Sc.D., F.R.S.

Regius Professor of Botany in the
University of Glasgow

Cambridge :
at the University Press

1912

CAMBRIDGE UNIVERSITY PRESS
Cambridge, New York, Melbourne, Madrid, Cape Town,
Singapore, São Paulo, Delhi, Tokyo, Mexico City

Cambridge University Press
The Edinburgh Building, Cambridge CB2 8RU, UK

Published in the United States of America by Cambridge University Press, New York

www.cambridge.org
Information on this title: www.cambridge.org/9781107672871

© Cambridge University Press 1911

First Edition 1911
Reprinted 1912
First paperback edition 2011

A catalogue record for this publication is available from the British Library

ISBN 978-1-107-67287-1 Paperback

Cambridge University Press has no responsibility for the persistence or
accuracy of URLs for external or third-party internet websites referred to in
this publication, and does not guarantee that any content on such websites is,
or will remain, accurate or appropriate.

*With the exception of the coat of arms at
the foot, the design on the title page is a
reproduction of one used by the earliest known
Cambridge printer John Siberch 1521*

CONTENTS

CHAPTER I

PRESENT-DAY BOTANY—A CONTRAST

FEW pursuits are more thoroughly misunderstood by the average person of education than that of the present-day botanist. It is not want of sympathy which leads to this. Almost everyone has an interest in the plants which he sees around him, sometimes from the point of view of their beauty, sometimes of their use to himself or to the human race at large : sometimes the interest is the more philosophical one of their place in organic nature, or of their origin in point of time as disclosed by the evidence of the fossils. On one or another of these grounds the botanist finds some feeling for the science of his choice already alive in the minds of his friends.

The conversation in ordinary educated society may often throw an illuminating light, revealing the layman's estimate of what a botanist is actually about. In this case it commonly appears that the estimate errs by being belated. Old time-worn aspects of the science are assumed to be still the

living problems of the day.   In no point does this
emerge more clearly than in the undue importance
attributed to the finding and recording of new
species and varieties.   This is a survival of the time
when the science, still in its infancy, was mainly
engaged in the recognition and tabulation of living
forms.   Personal credit was then apt to be measured
by the number of the new determinations.   Far be
it from me to suggest that the process of recording
new species is yet complete, or even approaching
completion.   But whereas in earlier periods most
botanists were engaged in this duty, the work is
now so far advanced that it remains in the hands of
comparatively few.   It is indeed a fact that some of
the most prominent investigators and writers have
never recognised or described a single new species.

An illustration of the misconception on this point
in the minds of well-educated people may be quoted
in my own case.   When preparing for a recent visit
to collect Ferns in Jamaica, a kindly Dignitary of the
Church expressed the hope that I should return with
several new species.   I, however, congratulate myself
on looking over my collections that all my specimens
fall under well-known and recorded determinations.
The reason for this callousness to novelties may be
demanded: it is this.   In the island of Jamaica the
Fern-Flora has been so well worked, that all the
most prominent and typical plants have been already

discovered and described. If new species were found, the chances are strongly against their presenting any feature of special importance for purposes of comparison with others: and in the absence of such the duty of describing and delineating is unprofitable for one whose aim is comparison with a view to tracing Descent. If this be the case for Jamaica, what will be the chance of discovery of important novelties in our own carefully searched islands? The home botanist can only look for fresh "finds" among minute and obscure forms, or among the subordinate shades of varietal distinction. To most of us the game is not worth the candle at home, and we do not lay ourselves out for it in the better known localities abroad. Specially organised expeditions in little known countries where much that is found is new, are a different matter; but they should be undertaken only by those specially trained for the purpose. Thus the recording of new forms no longer takes the premier place in the progress of the science.

If the conversation at some social gathering turns, as it often may when a botanist is present, to the floral decorations, the unfortunate victim of a misplaced confidence will as likely as not betray a lamentable ignorance of the most fashionable new varieties of decorative plants circulated from some celebrated nursery, or be unable to give the usual social or trade name of some common greenhouse plant;

though he might give an hour's lecture on its structure, and its systematic and biological characteristics. The sympathetic layman is disappointed and perplexed when he finds this, for he dearly loves a name. He does not realise that in point of fact names are necessary symbols: a means of recording, and nothing more: and that for the student the interest in a given plant may have begun long before the identification, and will as a rule intensify with study quite apart from the exact designation. It is when he puts his observations on public record that the necessity arises for accurate determination. The names which are used to stamp specific or varietal identities of plants are like the words of a language. Their value *per se* is small. It is in collocation that identities stamped by specific names acquire their interest and their worth. The individual who stores his mind merely with the names of plants may know as little of the science as one who memorises a dictionary would know of literature. It is considerations such as these that justify many botanists in their neglect to commit a multitude of names to memory. The knowledge of characteristics and of relations is the important matter, and this should go along with a facility in ascertaining the correct naming whenever it is required for purposes of description, in case the memory has failed to bear that burden.

Which of us has not been assumed to know at once by sight and to be able to name the various Conifers growing in the plantations round some country house, or planted proudly as "solo trees" in prominent spots to challenge the eye? The assumption is complimentary, but it is apt to be embarrassing. The plain fact is that the Coniferae are a family with an irritating sameness of habit for those who have not made them a special study, however distinct their features may appear to those who live with, and know so to speak personally, certain individual specimens. When one is found at fault in the specific distinctions of *Abies* or *Cupressus*, there is some risk of attainments in other branches being unduly discounted. Such knowledge is doubtless desirable, as is all knowledge: but to the majority it would be but so much mental ballast, and would take no direct part in the working of the intellectual ship.

Another subject which leads to misunderstanding is Horticulture. Theoretically every botanist ought to be an expert gardener, and doubtless every one of them would wish to be. It will usually be assumed that he is, but in many cases that assumption is ill founded. The man whose life-work is in the herbarium may have little time or opportunity, or it may be even inclination for horticulture. The laboratory student, even with a good knowledge of the current problems,

is often in the same position.   He will probably use
a microtome better than a trowel, and judge better
of the methods of fixing and embedding with a view
to cutting sections than of potting.   Such men will
probably have spent, perforce, the greater part of their
time in towns, and will thus have missed the experience
which falls naturally to those who live in the country,
and are observant.

Still another ground for misunderstanding is the
terminology of the science.   No subject has been
more heavily weighted by technical terms and uncouth
names than Botany.   The very pronunciation of them
is often an offence to the ear of the cultivated classic,
while to those who love nature and natural things
the language commonly used in Botany is an effectual
barrier to the pursuit of this spontaneous line of
interest.   A certain thoughtlessness of those within
the pale in the use of what can hardly be designated
otherwise than as their "slang" must be admitted.
It has given reasonable ground for the idea that a
botanist loves his terminology, and even glories in it.
As a matter of fact the profession groans under the
burden.   It is largely a legacy of a misguided past,
which can only be thrown off by a determined and
collective effort.   It is true old terms are constantly
becoming obsolete, and dropping out of use, but
others are being introduced to meet current needs.
The result is that the vocabulary to be heard at

any sitting of the Botanical Section of the British Association is certainly not such as is "commonly understanded of the people." The practical effect of this is an unhappy isolation, and the unconscious ostracising of many whose interests are already engaged in kindred questions. Occasionally among the exponents of the subject a brilliant exception arises, who either from the nature of his investigation, or from his power of conveying it, or better from the combination of both, succeeds in touching the general interest : and perhaps the most striking example was Charles Darwin, whose use of technical expressions in the *Origin of Species* was reduced to its lowest terms, a fact which conduced in no small degree to its effect upon the general reading public.

Several distinct causes of the misunderstanding between the modern botanist and the lay public have now been recognised. They all arise more or less directly from a common source. It is not generally realised how far the science has progressed in its differentiation. In the course of the last half century there has been a vast increase in the number of those who pursue it. Institutes in Universities and elsewhere have multiplied greatly, each with its more or less complete staff of workers. Most of these are contributing to knowledge by practical enquiry of some sort, and their observations and arguments are published in a continuously growing stream. Professor

Scott-Elliott in his *Botany of To-day* has estimated that about a quarter of a million pages of printed matter relating to botany are produced annually. Naturally it has become increasingly difficult for any one mind to grasp the multitudinous details, or to follow the descriptions of them as they appear. Accordingly it has become necessary for each one to specialise, if he is to be a practical worker at all : to take up some limited area of the science, and make it his own by reading and by personal observation. In proportion as this is realised other parts of the subject are apt to be neglected. This is the point which has not been fully grasped. It still remains to be learned that an expert on fossils of the coal may not ever have grown a living plant; an authority on physiology may be sadly lost in the determination of rare exotics; a leading cytologist may be hopelessly puzzled by the identification of the Conifers ; or a student of Algae may know little or nothing of the source and supply of condiments and drugs. And yet all of them pass under the comprehensive name of "botanist."

On these facts it may perhaps be hastily concluded that the position of the science is highly unsatisfactory : that each will take his own way, and that coherence of intention among the workers is at an end. It must be admitted that there is some danger of this. It is difficult even for the most enthusiastic student to keep himself adequately informed on the general progress of the science as a whole. And were it not for the compulsion which binds the professional botanist in most cases to his duty as a teacher, the risk would be greater than it actually is. Fortunately almost all have to take their turn both in elementary and advanced teaching of branches of the science quite apart from that which is the chosen speciality. This compulsion tends to right the balance, and leads to a periodical revision of the science, as it grows, by each teacher. Thus his interest is compulsorily kept alive over the general field of the subject.

In presence of this pronounced tendency to specialism, the amateur is apt to miss the main ends which the expert has ultimately in view. It would seem accordingly to be worth while to embody in a series of short essays some reflection of the outlook of an average botanist, himself a specialist, it may be, upon the ordinary objects that surround him. To show how he regards such vegetation as he would encounter on a holiday, and to touch lightly and with the least possible technicality upon some of the problems which arise in relation to them. Such lines of thought converge more or less directly towards one central problem, still so far from ulti-mate solution, viz., how the plant-organisms we see around us came to be such as they are, and where they are? The effort is in fact to reconstruct evolutionary history. It is clear when once this general problem is enunciated, its solution must involve enquiry not only as regards form and structure, but also as regards function with which form and structure are so closely related. These questions expand naturally into numerous cognate phases of enquiry, such in fact as the various individual specialists have made the subject of their detailed research. Thus, while apparently driving lonely furrows, they are all at work for a common end. But the field is so vast that the casual onlooker may fail to grasp the general scheme, by reason of visualising only that part of it which happens to be nearest to him. He may see perhaps a solitary investigator working in a way that may seem trivial or unsatisfactory. But the doubts of the critic may really arise from the fact that the value of the work cannot be estimated without some knowledge of its relations to the ultimate goal. On the other hand, it must be confessed that it is not every worker who understands the general scheme of which his labour forms a part, however well he may be carrying out his own particular research. And furthermore, since there is seldom any master-hand directing the efforts of individuals, some workers, having a perverted idea of the general scheme, may lay out their own investigations on mistaken lines, and their labour may be in vain. This is the fate which too often awaits the zealous but isolated amateur.

The fact is, regret it as we may, that the day of specialisation is upon us. The good old times, in many ways so sound and so stimulating, when one mind could compass a circle of the sciences and even contribute to them all, are gone for ever. Now it is more than the man of ordinary ability can do to keep himself even moderately informed over the whole area of a single science. Hardly any aspire to a special knowledge of more than one branch of the many into which each science is divided. It is useless to resist this state of affairs so naturally forced upon us. The important point is, however, to be fully aware of the weaknesses which are liable to follow in the wake of specialisation, and to use every opportunity to neutralise their results. The most effective safeguard is the proper grounding of every young student in a cognate group of the sciences before he is encouraged or even allowed to specialise. And those sciences should be so taught that he shall grasp thoroughly the lines of reasoning upon which their structure depends. So prepared, the young investigator may specialise with safety, the broad base of education being ready to support the superstructure of his exact, but localised, knowledge.

CHAPTER II

THE BEACH AND ROCKS

BOTANICALLY speaking the sandy Beach between the tide-marks is an uninteresting thing, for it is desert. You may walk for miles along it, and find below the high-tide mark no sign of active vegetation. Seaweeds may be littered here and there, but they will be in varying stages of decay. You take up a handful of sand and examine it: it will be chiefly composed of grains of silica, but with many fragments of shells of animals, and occasionally of the lime-incrusted remains of certain coralline seaweeds. But the conspicuous fact is that active vegetable life is absent from the moving sand. The reason for this is not far to seek. It is that, apart from the floating life of the open waters, or Plankton, as it is called, seaweeds with few exceptions require a fixed substratum. How different from the barrenness of the loose, disintegrated, and ever shifting sand is the condition of any rock or wooden pile which you may see exposed between the tidal limits, or extending below the low-tide mark far into the deeper water. Such fixed surfaces, except where they may be exposed to the grinding action of the sand washed past by the movement of the waves, are found covered by seaweeds of various forms and colours. These constitute a vegetation of Algae, widely different from that of the land. Their characters of outline, of function, and of propagation are peculiarly their own.

The spread of this characteristic Flora of the sea is limited upwards by the high-tide mark, and though some Algae appear content with a short immersion at high water, they are all dependent upon access to sea-water at intervals, while some require to be constantly submerged. Some straggle up estuaries and river-mouths till the brackish water is reached, but do not thrive in fresh water. On the other hand certain genera, such as *Batrachospermum* and *Lemanea,* allied to the Red Seaweeds by their structural and propagative characters, are common in freshwater streams but do not occur in the sea. These plants are green in colour, and might paradoxically be described as green freshwater red seaweeds. Passing through the zone of sea-weeds downwards to the low-tide mark and beyond it, it becomes a point of interest to enquire what are the limits of depth to which this Algal Flora may extend. It appears that at about 150 feet below the sea-level as a rule all growth ceases, though in some very transparent waters the limit may be deeper. It is in fact the fading out of the light necessary for the self-nutrition of these plants, by its absorption in passage through the great depth of water, that imposes a limit of extension upon the seaweeds downwards. And so we may understand that, exclusive of the floating Plankton, the character-istic Flora of the sea is restricted to a comparatively narrow zone lying between high-tide mark and some 150 feet below low water.

The plants which constitute this Algal Flora, so restricted in its spread, fall into three large groups, which are roughly characterised by their colour as the Green, Brown, and Red Seaweeds. Not that the mere tint is absolutely distinctive, as the already quoted case of *Lemanea* shows ; but it happens that the colours mentioned run in a measure parallel with other characteristics of form and propagative method by which the large groups are more strictly defined. It has been found that the colouring has its physio-logical meaning and importance in relation to the light so necessary for the self-nutrition of these plants. In the Brown and Red Seaweeds the greatest activity is carried on in the light of that part of the spec-trum which is complementary to their own colour. Ordinary green plants make special use of the rays in the red end of the spectrum, but for Brown and

Red Seaweeds rays further along the spectrum are specially effective, and it is such rays, in the direction of the blue end of the spectrum, which penetrate furthest into the depths of the sea-water. Thus the colourings have a probable relation to the nutrition of the Seaweeds which show them : and in particular, the possession of a brown or red tint makes self-nutrition at great depths possible, while it does not appear injuriously to affect life under less rigorous conditions.

From the circumstance that the Red and Brown Seaweeds are by their colouring specially able to use in their nutrition such light as penetrates far into the depths of sea-water, it might be expected that they would find their place exclusively at these low levels. But it is not so. There is no exact scale of zonation of the three types of seaweeds according to depth of water. Representatives of all may be found near to the high-tide mark. It is, on the other hand, an interesting fact that such full-green Algae as *Struvea*, which show no colour specialisation different from land-plants, may live at so great a depth as 150 feet. But speaking generally the Red Seaweeds are certainly more prevalent at the lower levels, where the green are present in less numbers. The Brown extend from the highest levels downwards, but stop short of the greater depths. As regards genera and species, however, there is more definite zonation, and it is a familiar matter of observation how the dense tassels of the yellowish *Pelvetia* almost take possession of the highest levels on our shores, followed downwards by definite zones of tangles, such as *Fucus platycarpus* and *Ascophyllum nodosum,* while the strange *Himanthalia,* with its thong-like fruiting branches rising from the cup-shaped base, are only met with below half-tide mark. Similar sequences of zonation may be noted in the Red Seaweeds also.

One point of distribution is specially worthy of remark: it is that certain of the Green Algae are stimulated to active growth by sewage impurities, which seem to affect the others adversely. In the neighbourhood of a harbour, and conspicuously near to the drainage-outfalls, the Green Seaweeds are in the ascendant: so much so that their presence, in quantity, may often serve as an indication and a warning. In recent years the growth of vast masses of the bright green Sea-Lettuce (*Ulva latissima*), and its subsequent decay as the season progresses, has been a cause of serious nuisance along the shores which border on the relatively impure waters of Belfast Lough and Dublin Bay. No method has yet been devised for checking its growth.

The Sea-Lettuce is common all round the British coasts, being found attached to rocks, piles and piers. It consists of a thin filmy expansion of

simple outline and full green colour, and it is fixed
by a narrow stalk which widens into a minute
attachment disc. The whole plant is very soft and

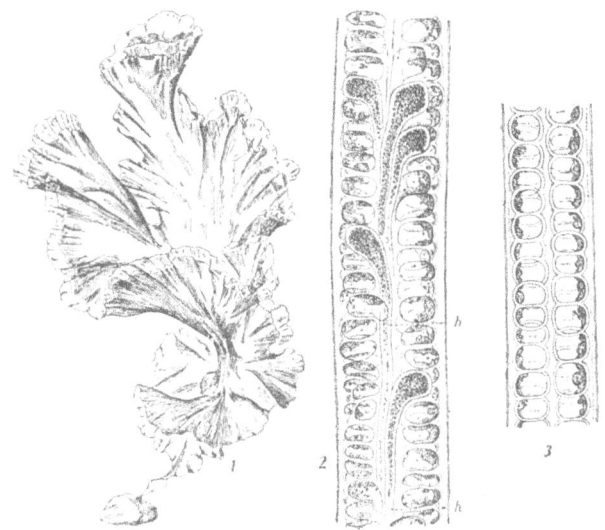

Fig. 1. *Ulva lactuca.* 1. A complete specimen of the plant.
3. A longitudinal section from the upper part of the flat expansion, showing the two layers of cells of which it is composed.
2. A similar section from the lower part, showing similarly the
two layers of cells; several of the latter have elongated into
tubes or hyphae (*h*). 2 and 3 are highly magnified. (From
Oltmanns' *Morphologie und Biologie der Algen*, after Thuret.)

B.   2

flexible, so that when left by the receding tide it
adheres to the substratum, but floats off again as
the water rises, yielding to every motion, but re-
maining fixed at the base during life (Fig. 1, 1).
Sections show that the flat expansion is composed
of two layers of oval cells, which are all alike,
having a protoplast containing a nucleus and green
chlorophyll-body, and surrounded by a swollen cell-
wall. But towards the base of the plant some of
the cells may be elongated downwards into tubes,
which force their way between the two layers, and
reaching the base add to the strength of the
attachment-disc (Fig. 1, 2, 3). Thus constructed the
plant may grow to a large size, though still pre-
serving its simple outline.

There are other genera of Green Algae, differing in
their form, but resembling *Ulva* in all essential points,
such as the common genus *Enteromorpha* with its
tubular body, gut-shaped as its name implies, produced
by the separation of its two layers which correspond
to those of *Ulva*; or *Monostroma*, in which at an
early stage the tube is ruptured, so that a broad
green expansion is produced, only one layer of cells
in thickness. But all such forms, including *Ulva*
itself, are held to be merely consequences of the
elaboration of the simple filament of cells, such as
is seen in the genus *Ulothrix*, various species of
which inhabit fresh or salt water (Fig. 2, *A*). If the

cells of a filament like that of *Ulothrix* be divided longitudinally as well as transversely, the transition to such forms as those just described would be easy: and judging from the fact that all these Algal types begin life as simple filaments, this conclusion seems to be a highly probable one. When it is noted further that the minute cells composing these Algae are essentially like those of *Ulothrix*, the mode of their origin can hardly be held to be any longer in doubt.

In plants at large a quiescent period of self-nutrition precedes propagation, and these Algae are no exception. It matters not whether it be a simple filamentous plant like *Ulothrix*, or a large expansion such as *Ulva*, during this first period it is capable of growth accompanied by cell-division to which there is no ostensible limit. The relatively large body thus produced, consisting of thousands of cells in such a plant as *Ulva*, is apparently inert, for the living protoplasts are enclosed and fettered each by its restricting cell-wall (Fig. 1, p. 17). The vegetative period is thus one of bodily quiescence, however great the physiological activity may be. But when in any matured and well-nourished plant the circumstances are favourable for propagation, a change may occur in individual cells, which leads to a period of very active movement. Their protoplasts, after sub-division, escape from the restricting cell-walls into

2—2

the surrounding water, each portion forming itself into a separate and actively motile body, or zooid. These minute naked bodies are slightly elongated or pear-shaped, while long cilia are attached to the narrower end of each, the active lashings of which give it rapid movement from place to place (Fig. 2, *C*).

There are in *Ulothrix* three different types of the zooids which thus escape into the water. Some are relatively large, and may be produced singly without division of the contents of the mother-cell, or more commonly by division of them into two or more parts according to the size of the mother-cell. They escape through an opening of the cell-wall into the water (Fig. 2, *B*). Each has four motile cilia attached to its narrower end. A period of active movement in water ensues, varying in length according to circumstances. In this way these zoospores may scatter in different directions, and travel a relatively considerable distance. They finally settle upon some solid substratum, forming each a new cell-wall and grow out transversely to their former axis. A rhizoid of attachment is put out in one direction, and in the other there grows a new filament like the original parent. There is thus a means both of increase of numbers and of dispersion of the new individuals. Other similar but rather smaller zooids have been distinguished as micro-zoospores. They are of the same type as the first, but are produced in larger numbers in each parent-cell. In size they resemble the third type, which are the sexual zooids,

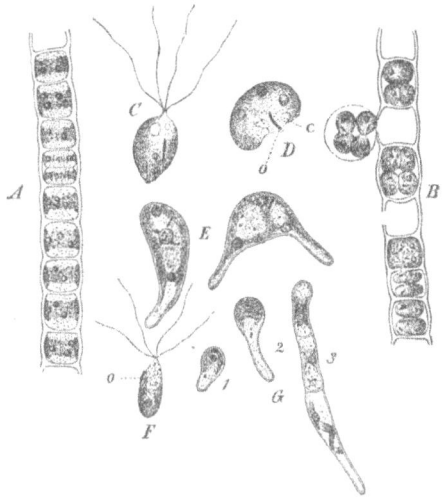

Fig. 2. *Ulothrix zonata*, showing formation of zoospores. A=A filament in the vegetative condition. B=a filament whose cells are producing zoospores. C=a single zoospore. D=a zoospore coming to rest: o=eye-spot: c=pulsating vacuole. E=germinating zoospores. F=micro-zoospore. G 1—3=germinating microspores. A, B × 200. C—G 1, 2 × 333. G 3 × 166. (After Klebs.)

or gametes; but these are readily distinguished from them by the number of their cilia, which are two in

place of four. They differ also in their behaviour,
for it has been observed that the gametes which
escape from different filaments when they meet
coalesce in pairs (Fig. 3 *A—D*). The result of the
fusion is a cell called a zygote (*E*). It soon loses its
cilia, takes a rounded outline, and forms a cell-wall (*F*).
After a period of rest the zygote germinates, bursting
its cell-wall, and the contents divide into four parts
(*F*, *G*), each of which may grow into a new filament
like the parent. The propagative methods in *Ulva*
and other related Algae are similar to those in *Ulothrix*, but they have not been followed out with the
same exactitude. It has been shown by Professor
Klebs that the various forms of the motile propagative
cells do not follow in any necessary sequence. It
would seem probable that their differences are cor-
related with external conditions, and that it is these
which decide in what sequence they shall appear,
rather than any internal necessity.

Such a story of propagative method as this in
Ulothrix, combined with the similar facts which
result from the study of other allied forms, leads
to various general reflections. In the first place it is
seen that, in the cycle of life of *Ulothrix* or of *Ulva*,
two phases of existence appear. The one is stationary,
a condition imposed upon it by the fact that the living
protoplasm of each cell is encysted, that is confined
within a cell-wall, which makes free movement

impossible.   This is the phase commonly known and recognised as *the plant,* which by growth and repeated division of its cells may attain large size.   The other phase is one of freedom and active movement.   The

Fig. 3.   *Ulothrix zonata.* Showing production of gametes. $A=$ filament from which the gametes (sexual cells) are escaping. $B=$a gamete. C, $D=$conjugation of gametes, in pairs. $E=$ zygospore. $F=$zygote, or result of fusion. $G=$germs produced from it. $H=$a filament with germinating zoospores (z). A, H, F and $G\times200$. B—$E\times333$.   (After Klebs.)

protoplasm after division escapes from the restricting cell-wall in the form of zooids, each capable of independent movement in the water into which they escape.   From their minute size and their dependence upon suitable external conditions for their escape and

movement they are apt to be overlooked.   But there
is no doubt that there is a regularly recurrent phase
of motility in every completed life-cycle.   It may
then be enquired which of these two phases, the
stationary plant or the motile zooid, was probably
the prior condition in evolution.

Among certain simple aquatic organisms called the
Flagellates minute forms exist which are always freely
motile, and have no fixed and encysted stage.   Each
individual consists of a naked protoplast similar in
essentials to the zooid of *Ulva* or *Ulothrix*.   This fact
suggests very strongly that the free and motile
condition was primitive, and that the encysted state
with a cell-wall surrounding each stationary protoplast
was later and derivative.   The individual cells of the
Ulva plant are so minute that they are not seen
individually with the naked eye.   But the obvious
Ulva plant is a large congeries of them.   The cells
of the Flagellates are also individually small, but
being isolated they are invisible to the naked eye.
And so it is that while an ordinary visitor to the
sea-shore may be familiar with so obvious a plant as
Ulva, he will be unaware of the existence of its
propagative cells, as well as of its simpler unicellular
prototypes the Flagellates.   This was for long the
position of Algologists also, and the view above
stated has emerged only in comparatively recent
times.   The suggestion thus made for *Ulva* or

Ulothrix may be extended to other plants as well. It may be stated as a general hypothesis that primitive life was aquatic, and certainly one of its forms, if not indeed a general form of it, was free-swimming. The quiescent stage with encysted protoplasts characteristic of the vast majority of plants now living was probably derivative in many cases, and perhaps in all, and owed its origin to the convenience which it offers for protection, as well as for self-nutrition and growth. If this be true, then it is the derivative condition which we commonly recognise as the "plant."

A second point for consideration relates to the origin of sexuality. It has been seen that two similar gametes unite in *Ulothrix* to form the zygote. There is no differentiation of sex, though the fusion is of the same nature as where such a distinction exists. It would thus appear that some process of fusion of cells antedated in descent the distinction of sex. Further, the general similarity in origin and form of the gametes and the zoospores cannot escape remark. It would appear as though the gametes were attenuated zoospores, a view which accords with their origin by further subdivision from parent cells similar to those which form the zoospores. It is possible that the fusion in such forms as these originated as a means of strengthening the attenuated zoospores. Again, comparison with the Flagellates would support this hypothesis; for in these plants such fusions are absent, while the plant itself is here of the nature of a zoospore. It thus appears probable that the motile zoospore condition was prior to such cell-fusions as the gametes undergo.

Thirdly, a comparison of *Ulva* or *Ulothrix* with other Green Algae, shows that inequality of size of the gametes, and ultimately marked differentiation of the sexes, may appear. Such a progression from equality to inequality of the gametes is seen also in various types of the Brown Algae. From such facts it may be concluded that the differentiation of sex was acquired subsequently to the initiation of the process of fusion, and that it was attained in several different lines of descent. Its cause is probably to be found in the advantage which a larger store of nutriment brings to the germ. One type of gamete, the male, retained its motility and approximately its original size: the other, the female, became enlarged and receptive, and ultimately it sacrificed its motility to the advantage of better nourishment of the young individual on germination.

It would be impossible here to do more than suggest such views. The full discussion of them must be left to larger works. But it will be apparent that a study of the common *Ulva* and its related forms, together with a comparison of other Algae higher and lower in the scale, raises questions which are of prime importance as they affect the outlook upon the vegetation of the earth. Guided by such study we shall regard the condition of simple organisms motile in water as primary ; the encysted state of the protoplasm and the fixed position as a secondary condition acquired as a concession to nutritive convenience and mechanical protection ; the process of sexual fusion itself appears also to be secondary, originating perhaps as a method of strengthening attenuated zoospores ; while, lastly, the sexes were in the first instance alike, the differentiation of the sexes having arisen subsequently to the origin of sexuality. It probably arose in relation to the necessity for adequate nutrition of the germ, so that after fertilisation it may make an efficient start in life, and successfully perpetuate its race.

Returning in conclusion to the Flora of the shore as a whole, with its varied forms, Green, Brown, and Red, they together with certain dwellers in fresh water form a vegetation apart. These plants have developed in accordance with their aquatic surroundings, and in their own way some of them have attained a high complexity not only of form and construction, but also of propagative method. But they always bear the impress of their aquatic habitat, and differ in essential features from the characteristic Flora of the land. Their higher representatives may then be regarded as being the culmination of their evolutionary history. They are in fact "blind" branches of the evolutionary tree. So far as any relation can be traced between the Algae and characteristic Land-plants this is to be sought for among the simpler rather than the advanced types, and even then the indications are of the slenderest. But notwithstanding this, it is from the Algae that we receive the clearest indications how plants at large attained to many of the most fundamental features that they show. The persistence of some of these characters among Land-plants, even when their presence is inconvenient, makes the conclusion inevitable that the origin of the vegetation of the earth was ultimately aquatic. It probably sprang in the first instance from free-swimming plants such as the Flagellates, while the fixed Algae of our shores represent an intermediate state, in which the quiescent phase of self-nutrition had already become a constant, as it is in them an obvious and indeed a preponderant feature.

CHAPTER III

THE BRACKEN FERN

A few miles along our coast there is a headland backed by woods which stop short before the coast-line is reached. The vegetation that runs down to the beach is largely composed of the plants usual on sandy shores, but they are accompanied by a quantity of Bracken Fern which extends down almost to the high-tide mark. Here and there the Bracken may be so thick as almost to exclude any other plants. It spreads backwards also into the woods, but there various other Ferns make their appearance, and these with the Bracken constitute a chief factor in the undergrowth below the canopy of trees.

There are few plants that are so widely spread over the earth's surface as the Bracken. It is the most prevalent Fern here at home and is common through Europe. It is seen on the high levels of Ceylon, and has been noted as specially prevalent at Singapore. It grows on the West Indian islands, and though it is only sparingly present in South

America, it extends through the Southern Hemisphere to New Zealand. It is true that in these distant lands the type may differ from our own Bracken in minor details, but the differences are held to be only varietal, and the species designated *Pteridium aquilinum*—or better known under its old name of *Pteris aquilina*—may properly be recognised as cosmopolitan. Where it occurs it is commonly found in quantity, indeed it has been described by Dr Christ as the gregarious Fern *par excellence*. And so we see it sometimes growing with us so densely as to occupy the ground to the exclusion of all else. This is notably the case where the soil is sandy, and free from lime.

Wading through Bracken as high as one's waist it may be a surprise to be told that all that we see above ground is the leafage, and that the stem which bears the leaves is entirely concealed. But if you take the trouble to dig the plant up you will find some inches below the surface of the soil an elongated, horizontally running stock about half an inch in diameter or less, dark brown in colour, and marked laterally by two paler lines. From this the leaves arise alternately on either side (Fig. 4). The growing tip of the stock is covered with a dense mat of hairs, and near to it one or two young leaves may be seen. Further from the tip there will be found the stalks of one, or at most two leaves of the current season whose

blades are visible above ground. Further back again
are the stumps of the leaves of earlier seasons, the
upper parts of them having decayed. The stock gives
rise to numerous roots, which are dark and wiry, and
extend in all directions. The whole shape of the
plant appears at first sight to differ strongly from

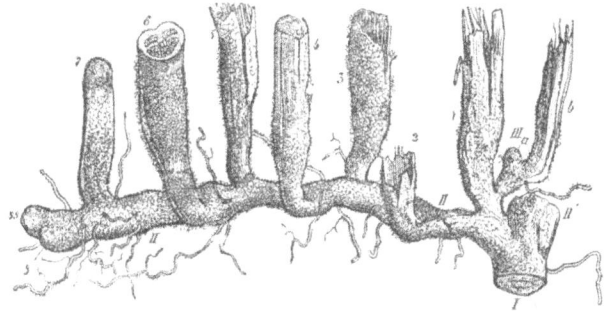

Fig. 4. Rhizome of Bracken Fern, showing the bases of successive
leaf-stalks (1—8): No. 5 is the leaf of the preceding season,
6 that of the current year, cut so as to show "King Charles'
Oak," 7 that of the coming season, and 8 a still younger one.
ss is the apex of the stock. Roots run out in various directions
from the rhizome. (From Goebel, after Sachs.)

that of the densely tufted Ferns such as the Shield-
Fern (*Nephrodium Filix-mas*); but intermediate
conditions may be found between the two types of
shoot, which show that the difference depends merely
upon the degree of elongation of the axis : where

this is short as in the Shield-Fern the leaves appear as a compact tuft, where it is long they are isolated, while the stem may be either exposed to the air or it may run underground as in *Pteridium.*

An examination of a number of rhizomes will show that the stem is capable of indefinite growth in length, but occasionally it forks at the tip into two exactly equal branches. It is plain that by this means the shoot-system may be enlarged, but a more effective way is that which is more common in the Bracken, viz. by the formation of a new bud near the base of each leaf-stalk. Since each such bud is potentially a new shoot endowed with apical growth like the original, it is apparent that there are ample powers of extension of the shoot-system. On the other hand, if a stem be followed backwards from its apex, sooner or later a region will be reached where the tissues will be no longer fresh and living. The fact is that decay is constantly progressing from the base upwards, and whenever it extends beyond a bifurcation or the insertion of one of the leaf-buds there will be a complete severance of the two living apical regions, and two individuals, physiologically independent, will result. In this way the Bracken may spread over large areas of ground, its development being simply vegetative. In fact those large sheets of its growth which occupy wide stretches of hillside and woodland owe their origin to continued growth of individuals originally few in number. The underground position of the stock makes for permanence of the individual. During the austere time of our northern winter, or it may be the drought of a southern summer, it contains a liberal store of food-material laid up in its fleshy tissues, ready for use in forming fresh leaves in the coming season. This habit is naturally very effective against extremes of climate. It has probably been a factor in the success of the Bracken, as shown by its commonness at home, as well as in its cosmopolitan spread elsewhere.

The stately leaf of the Bracken is borne by the upright leaf-stalk, which is dark coloured and hairy where buried in the soil, but green and smooth above, while paler lateral lines follow its course upwards. It is here that provision is made for gaseous interchange with the outer air, which is elsewhere prevented by a covering of hard woody tissue. It is this leaf-stalk that children cut obliquely with a knife, and see on the cut surface what they call "King Charles' Oak" : the complicated arrangement of the conducting strands and of the brown tissue that accompanies them being in outline roughly like a tree. In Germany it is held to resemble the Double Eagle, and the name "Adlerfarn" is accordingly given to it. High up on this rachis the lateral lobes, or pinnae, are borne in pairs, diminishing successively in size till the apex is reached. The pinnae may be again branched, and the branching may be further repeated, so that the whole leaf takes a highly divided form. The ultimate ramifications widen out into green expansions which collectively make up the lamina or region of the leaf effective for self-nourishment. So well is this function carried out that a large overplus of the material gained is stored in the fleshy tissues of the underground stock, and carried over as a physiological balance to the ensuing season.

It is no wonder that a plant having the advantage of a well-developed stock for storage, deeply buried in the ground and so protected against risks from climate as well as from animal attack, should succeed. But in many districts, and especially in the North of Scotland, its success is embarrassing to the farmer who wishes to graze sheep, or to the forester who has to plant out a hillside. To them the question is how to extirpate it cheaply and effectively. The method in common practice is physiologically sound. It is to mow down the leaves about the end of June, that is so soon as the leafage is approaching full development. The process is to be repeated a second, or if necessary a third year. Some may think that this succeeds because the propagative spores, which as we shall see are borne upon the leaves, are checked in their development, and so their germination is prevented. But it will be pointed out that the growth of new plants through the germination of spores is uncommon

Fig. 5. Leaf of the male Shield-Fern (*Nephrodium filix-mas*, Rich.), about one-eighth natural size, the lower part having the under surface exposed. To the left a single segment, bearing the sori, each covered by its kidney-shaped indusium. (After Luerssen.)

3—2

on the open hillside.   The explanation of the success
of mowing about midsummer day is probably to be
found in its effect on nutrition.   The plant has at that
time of year devoted all, or at least most of its
floating capital of stored food, to the business of the
formation of leaves and spores.   But the leaves are
the nourishing organs, and they are mown before
their function is fully started.   Thus the plant loses
the capital it has embarked.   There may be a sufficient
residuum of stored food left to produce a second
or even a third formation of leaves ; but repeated
mowings will so deplete the rhizome as to lead finally
to death by starvation.

An observant friend in the Highlands has pointed
out to me that mowing is more effective at low levels
where the growth is strong, than higher on the hills
where the Fern is relatively stunted.   This is probably
to be explained by the fact that where growth of the
leafage of the year is favoured, the plant draws more
deeply on its floating capital than where the circumstances are less encouraging.   The hill plants will
probably retain a larger proportion of their supply in
the underground stock, and so be better able to renew
the effort at leaf-formation.

But nutrition is not the only function of the
Bracken leaf. In fully matured plants the leaf also bears
the organs of propagation.   If the ultimate branch-
ings of the leaf be carefully examined, it will be found

that in many or even in most of them the margins are curved downwards, and extended into a pellucid veil or indusium, so that what appears from above to

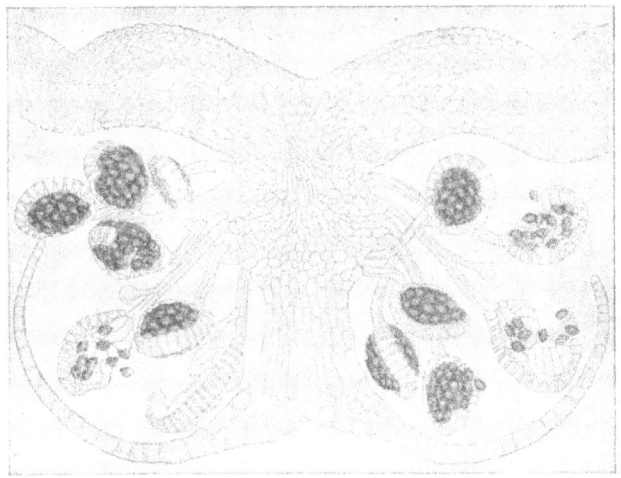

Fig. 6.   Vertical section through a fertile leaf, of *Nephrodium* traversing a single sorus.   This is attached to one of the veins of the leaf which is enlarged as the receptacle.   Upon it are seated the numerous stalked sporangia, each containing many dark-walled spores.   The whole is covered over by the umbrella-like indusium.   (After Kny.)

be the margin is really a curved and slightly swollen region extending into a continuous protective flap.

It is in the hollow of the curve thus formed that the sporangia are inserted in large numbers, while a second flap, less obvious than the first, forms a further protection, the sporangia lying between the two. A somewhat similar arrangement is seen in the common Shield-Fern (*Nephrodium filix-mas*), but here the sori, each covered by a kidney-shaped indusium, are borne on the under surface of the leaf (Fig. 5). Each of the sporangia is a stalked body with a capsular head shaped like a biconvex lens (Fig. 6). Its margin is occupied for three quarters of its length by a row of hard cells forming the annulus, or spring by means of which when ripe the sporangium opens and flicks away its contents. It is inside these capsules that the spores are produced, to the number of 48 in each. When ripe they are dry and dusty, and being very minute they may be blown to a distance by any breeze. If a mature leaf be laid to dry face downwards upon a sheet of paper, a very perfect print of its outline will in a short time be marked off by the spores which it will shed. Each spore is so small as to be invisible to the naked eye, but they are conspicuous collectively owing to their vast numbers. A moment's consideration is enough to realise the immense powers of propagation with which the plant is endowed. Each sporangium bears about half a hundred spores : each minute subdivision of the leaf may produce very many sporangia : the whole leaf is built up of a vast number of such subdivisions, and the tracts of Bracken may cover acres of ground. Calculations based upon such facts soon reach figures which it is beyond the powers of the mind to grasp : it must suffice to say that the spores are produced in myriads,

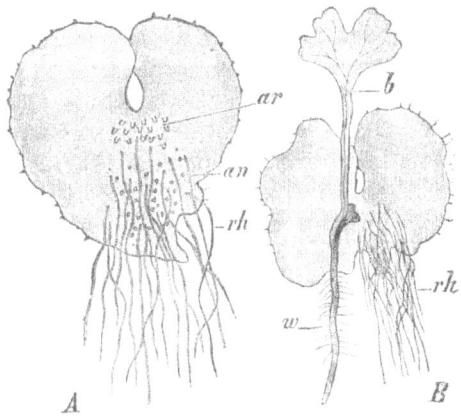

Fig. 7.   *Nephrodium filix-mas.*   *A* = prothallus seen from below. *ar* = archegonia.   *an* = antheridia.   *rh* = rhizoids.   *B* = prothallus with young Fern attached to it by its foot.   *b* = the first leaf. *w* = the primary root.   (× circa 8.)   (After Strasburger.)

while each spore though consisting only of a single cell bears all the powers and possibilities of a potential life.

It is well to note that the spores are matured upon leaves fully exposed to the air, and that dry circum-

stances favour their shedding. Moreover the Bracken
plant itself is well able to stand even considerable
drought. Thus not only is it an effective and successful plant of the land, but dry conditions favour the
dissemination of the spores, the production of which
is the final office of the plant.

But such dry conditions do not suffice for the
further development of the spores. Nor do they
germinate directly into new Fern-plants, but into a
body called the prothallus, differing in texture and
in other qualities from the parent. Moisture and a
suitable temperature are necessary for its growth,
and it finally forms a flat green scale-like body, about
half an inch or less in diameter, and attached by
delicate root-hairs to the damp soil on which it grows
(Fig. 7, A). It is a physiologically independent body
capable of self-nourishment, but usually of limited
growth and duration. Its delicate texture makes
continued moisture a necessity for its growth, and
consequently it is found only in damp and shady
spots, where spores happen to have been carried as
loose dust by the breeze. This circumstance goes
far to determine the positions which Ferns habitually
hold in Nature, for as we shall see the Fern-plant
springs eventually from the minute green scale-like
prothallus. Sooner or later it bears the sexual organs
upon its downward surface, the male or antheridia
nearer the base, the female or archegonia nearer the

indented apex. A succession of these is produced, so that the possibility of carrying out their functions may be spread over a considerable period.

When mature their condition is such that on access of external fluid water, as when rain falls, both of these organs, if ripe, burst. Naturally the lower surface of the prothallus on which they are

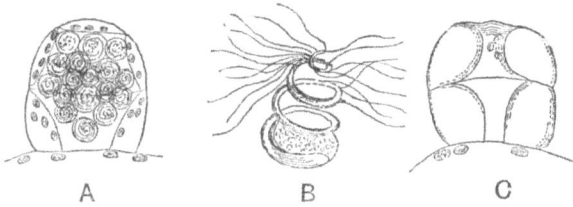

Fig. 8. *A.* An antheridium of a Fern, mature and containing numerous sperm-cells. *C.* A similar antheridium after rupture and escape of the spermatozoids. *B.* A spermatozoid more highly magnified, showing its spiral form during its movements in water by means of the numerous thread-like cilia. (From F. Darwin's *Elements of Botany.*)

borne will be moistened by any shower that wets the soil thoroughly, and the rupture of such as are ripe ensues. The male organs discharge their contents as spermatozoids into the water, in which they are capable of moving their spirally curled bodies rapidly by means of the active lashings of numerous cilia (Fig. 8). The female organs under similar circumstances burst at the projecting tip (Fig. 9, *B*), and an open channel then leads down to the egg, which is deeply seated in the tissue (Fig. 9, *C*). It might at first sight appear a matter of remote chance that any individual spermatozoid should enter an archegonium, but this is what may be seen to occur with a high degree of certainty. The explanation lies in the fact that the archegonium gives out by diffusion into the water a very small quantity of a soluble substance which acts as an attraction to the spermatozoids. They move actively from the weaker to the stronger solution, that is to the neck of the archegonium, which is the centre of the diffusion. One or more of them may enter the neck, and passing down it one may be seen to fuse with the egg-cell. This is the act of fertilisation, which consists in the fusion of two cells, and especially of their two nuclei ; for notwithstanding its strange form the spermatozoid is actually a cell, bearing a nucleus which forms the greater part of its spiral body. The result of the fusion is a cell called the zygote. It soon surrounds itself with a cell-wall, and forms the starting point for the development of a new Fern-plant similar to the original parent. The embryo plant is nursed for a time by the prothallus, upon which it lives as a parasite while young. But soon it becomes independent, rooting itself in the soil, and expanding its young leaves to carry on the function of self-nourishment.

It is very beautiful to note in young plants raised by artificial culture how the underground habit of the Bracken is started. The young seedling is in its first stages like any ordinary Fern: but after the formation of half a dozen or so of leaves the

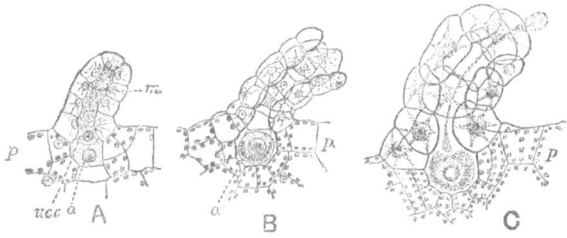

Fig. 9. *A.* An archegonium of a Fern, approaching maturity. The protective neck is closed at the apex. *m* = canal-cell. *v.c.c.* = ventral canal-cell. *o* = ovum. *B* and *C*, similar archegonia quite mature. The apex of the neck has ruptured, in both cases, and the ovum (*o*) remains embedded in the prothallus (*p*), while the open channel of the neck leads down directly to it. (From F. Darwin's *Elements.*)

axis bifurcates, and the two equal branches that are formed curve strongly downwards, and bury themselves in the soil. They never emerge again from it in their normal life. Together with the stores of nourishment laid up in their fleshy tissues they are fully protected against extremes of climate, as well as from animal attack.

The special efficiency in the Bracken is no doubt in large measure due to this habit. It has been seen how readily it spreads by the growth and profuse branching of the underground rhizome. This serves as a means of perennation from season to season. It also serves as a place of storage of reserve food for future use. So successful are these means of perpetuation and of spread that little dependence need be made upon the formation of new plants by germination of the spores and sexuality. There is many an observant botanist who has never seen a young seedling of *Pteridium* in the open, and we can only conclude that in Nature the completion of the life-story by sexuality is a comparatively rare event. There can be no doubt, however, that the power of raising a young plant through the germination of every matured spore is present, for under conditions of suitable culture the prothallus is easily grown, and young plants follow.

And thus the cycle of life of the Fern is completed. It is plain that it consists of two distinct phases, or generations as they are called, and the life-cycle is punctuated by two periods when the organism consists of only a single cell, viz. the spore which is produced by the Fern-plant, and the zygote which is produced by the prothallus. Between these landmarks intervene two more extensive developments or generations, viz. the prothallus and the Fern-plant. These succeed one another in a regular alternation. A produces B, and then B again produces A : and so on. These two generations differ widely in form and texture, as also they differ in their water-relation. The Fern-plant is structurally a land-growing plant, and is capable of undergoing free exposure to ordinary atmospheric conditions. Moreover dryness is essential for the final end of its existence viz. the shedding of its spores. On the other hand the prothallus is structurally ill fitted for exposure to dry air, while fertilisation, which is the end of its existence, can be achieved only in the presence of external fluid water.

Fertilisation may rightly be held to be the most critical event in the whole life of the plant. . It has doubtless been a phenomenon which has recurred in every completed cycle of life of Ferns throughout their long course of evolution. Its absolute dependence upon the presence of water is then a fact that cannot be lightly put aside. The same is also found in all such plants as Liverworts, Mosses, Horsetails, and Club-Mosses, as well as in the Ferns. Now the plants named constitute a Land-Vegetation which, on grounds of comparison as well as on the evidence of the geological record, is rightly held to be more primitive, and to be lower in the scale of evolution than the Seed-bearing plants. They occupy a middle place between these and the Algae. The latter are typically aquatic, and their fertilisation is usually carried out through the medium of the water in which they live. We have seen examples of this in *Ulothrix* and its allies. It therefore seems a probable interpretation of the facts that long ago certain forms of Algae spread from the water to the land; that they retained signs of their aquatic ancestry both in their method of fertilisation and in some measure as regards the texture of the prothallus. Since the final duty of the prothallus is the formation of the sexual cells, or gametes, it is called the Gametophyte. A second generation or phase, represented by the Fern-plant, underwent special development to fit it for more extended life exposed to the air on land. It has as its final result the production of spores, and it is accordingly called the Sporophyte. A Fern is then an organism, so to speak, with one foot on land, the other in the water. It indicates in the successive events of its life-history a past migration from life in water to life upon land-surfaces. The gametophyte is the conservative phase which retains its aquatic characters, the sporophyte is the innovation, which has assumed characters more suitable to the adopted habitat. Truly the life of such plants may be described as amphibious.

We have seen that in the Fern the prothallus is relatively small, while the Fern-plant is relatively large. As we rise in the scale of vegetation to the

Seed-plants, the same two phases are still found to be represented in the life-story, but their disparity becomes still more marked. Moreover the prothallus in the higher plants no longer leads an independent existence, but is enveloped in the tissues of the sporophyte, upon which it lives as a sort of parasite. As in other parasites the nutritive system, not being necessary, is reduced. This reduction of the gameto-phyte in the highest Flowering Plants attains to a high degree. All that can be held to represent the gametophyte in them is a body consisting of a few rudimentary cells, which produce or minister to the sexual gametes. Without pursuing these com-parisons further, we may sum up the essential points in one sentence and say that, in the higher Flowering plants the sporophyte is dominant, the gametophyte is evanescent. In them the sporophyte generation is what constitutes the obvious Plant-Life of the land, while the gametophyte is so inconspicuous that to the lay public it is a thing unknown.

The Fern-plant, though well able to hold its own as a dweller on dry land, is after all an ill-differ-entiated thing. Its primitive character is seen from the fact that the leaf serves various purposes. Most obviously it is an organ of nutrition as well as of propagation. But as we rise in the scale of organisation to the higher Seed-bearing plants these two functions become separated, and localised in distinct regions of the plant. In fact the Flower comes into existence as a thing distinct from the merely vegetative shoot. The discussion of this will be the subject of the ensuing chapters, in which it will be shown that with the appearance of the Flower the amphibious life is left behind, and a true Flora of the Land is brought into being.

CHAPTER IV

THE FLOWER AND METAMORPHOSIS

THE end aimed at in the average flower-garden is a succession of blooms as continuous as possible the year through. Even in the open air it may be managed that the garden shall never be actually without them, though naturally the barest time is in the dead of winter. The amateur gardener must not only admire, but must also study to produce this result; and in doing so he will realise much more fully than the casual lover of flowers that a period of vegetative activity is a necessary prelude to that of flowering. He will know that the very materials of which flowers are made must be laboriously acquired in all ordinary plants by self-nutrition, for which the green foliage is essential. In the annual plants of the garden this is obvious enough. Germination of the seeds of annuals in spring leads to the establishment of the leafy plant, and it is only after this has attained a certain development that flower buds make their appearance, and finally the expanded blooms. These are formed at the expense of the materials which the co-operation of the foliage-shoot with the root-system has been able to produce under the influence of sunlight. But in the case of many of the plants that expand their flowers in the winter or early spring the matter is not so simple. We see the Snowdrop or Crocus emerge from the soil, and we are apt to forget the swollen underground parts from which they sprang, and from whose stores they drew their sustenance. Still more are we unmindful of the coarse leafage which, following on the flowers of the preceding year, produced by its activity the very materials which we see in the spring worked up into the flowers that we prize. The fact that those materials are stored up out of sight so long before in the buried bulb or corm breaks the con-tinuity of observation, and provides the excuse for forgetting their existence. It is needless to elaborate illustrations of the simple fact that nutrition must precede propagation. However indirect the sequence of events may seem, some form of vegetative shoot precedes the flower with invariable constancy in the individual life of the higher plants.

As in the preceding paragraph so in ordinary conversation we distinguish the flower from the foliage-shoot. But if we examine them both we find that they have many features in common. In both there is a stem, which bears appendages. More-

over the stem of the leafy shoot may be continued
directly upwards into the flower. Here, however, it
abruptly terminates, while in the leafy shoot it may
continue to grow indefinitely at the tip. The result
of this is that the flower is borne at the end of its
stalk. The relation of the appendages or leaves to
this stem is similar in the foliage-shoot and the
flower. In either case they arise laterally upon it,
and their succession is such that the oldest are
lowest down, and the youngest nearest to the tip.
In many cases the appendages are developed equally
all round the stem, giving what is called a radial form
to the shoot or flower, and this is believed to be
a relatively primitive condition. But both foliage-
shoots and flowers may be unequally developed, the
appendages growing stronger on one side than on the
other. This results in lopsided forms, and it can often
be shown that their lopsidedness originated as an
adaptation to meet special needs of the organism:
in fact it is regarded as a secondary modification in
the history of descent. The flowers of a Pea or an
Orchid will serve as illustrations of this. In these
cases the adaptation has been to fit them for the
insect visits which are essential for the process of
cross-pollination. Lastly, the parts of the flower are
as a rule closely aggregated together, while those of
the vegetative shoot may commonly be separated by
intervals of the elongated stem. But all the leafy

shoots terminate in a compact aggregate of leaves, which is called a bud : and so, at least in the young state, the vegetative shoot may resemble the flower so far as the aggregation of its parts is concerned.

It thus appears that there are marked analogies between the foliage-shoot and the flower. This leads to the question whether there is any absolute difference between them. It will lie near to the hand to suggest the qualities of texture of the parts, of their colour, or their scent, by which flowers are apt to be characterised as distinct from vegetative shoots. But none of these are constant features in the one or entirely absent from the other. There is, however, one absolutely distinctive character. It is the presence in the flower of the organs of propagation called "sporangia." These are of two sorts, on the one hand the ovules which mature into the seeds, on the other the pollen-sacs which produce the pollen. It is the presence of either or both of these which stamps the flower as distinct from a foliage-shoot (Fig. 10).

However different these two essential organs of the flower, the ovule and the pollen-sac, may seem to be to day, there is good reason for holding them to be modifications produced in descent from a single primitive type of sporangium. This organ had as its office the production of spores or separate germs shed from the parent. In the more primitive plants these were all alike, as were also the sporangia which contained them. Such a condition is still seen in the Ferns, Club-Mosses, and Horsetails of the present day. It is held that sporangia essentially similar to theirs were modified in the course of descent, in accordance with the differentiation of sex,

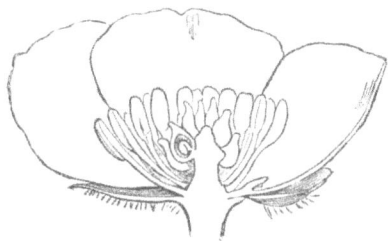

Fig. 10. Median section of the flower of a Buttercup, showing its constituent parts. On the outside (lowest down in the drawing and shaded) are the sepals of the calyx: within this the large petals of the corolla of which three are shown: within this and seated higher on the axis are the numerous club-shaped stamens, each of which bears four pollen-sacs. Centrally in the flower are the numerous carpels, one of which is dissected so as to show its single ovule, or future seed. (From Le Maout and Decaisne.)

until they finally assumed those two apparently different forms, the pollen-sac and the ovule. It is important for our enquiry into the relation of the flower and the foliage-shoot that this correspondence of the ovule and the pollen-sac with the sporangia of Ferns and Club-Mosses should be realised. Its acceptance makes it possible to compare the regions of the plants in which those organs occur ; that is, to compare the flower of the higher plants with the fertile region of Ferns or Club-Mosses. It is upon such comparisons that it will be found possible to base an opinion as to the true relation of the flower to the foliage region, and to understand the underlying similarity of construction which they show.

The obvious analogies which exist between the flower and the foliage-shoot in the higher plants did not fail to attract the attention of some of the early observers. Wolff, in his *Teoria generationis,* published in 1759, wrote that "In the whole plant whose parts we wonder at as being at the first glance so extra-ordinarily diverse, I finally perceive after mature consideration and recognise nothing beyond stem and leaves....Consequently all the parts of the plant except the stem are modified leaves." Not many years later the poet Goethe elaborated this general conclusion into the theory of "Metamorphosis," which is associated with his name. He recognised as Metamorphosis the process by which one and the same organ, for instance the leaf, presents itself to us in various modifications, such as the foliage leaf, sepal, petal, or stamen. He then went on to distinguish as examples of "progressive metamorphosis" those changes of type of the appendages which proceed from the cotyledons or seed-leaves through the foliage region and the bracts to the flower, and finally to the perfected fruit. He noted the modification of one form into another in a regular succession, till the "acme of Nature" is reached in reproduction by seed. On the other hand he designated as "retrogressive metamorphosis" the process by which that succession appears to be reversed, as for instance in abnormal or doubled flowers, when a stamen or a carpel developes as a petal, or even as a foliage leaf. Lastly, he styled as "irregular or accidental metamorphoses" those changes of the normal parts which often follow upon the attacks of insects, or other external agents. These general ideas of the relation of the vegetative and floral regions were amplified and made more definite by subsequent writers, and were for a long time widely held. Thus it became a general belief that the flower had resulted from changes wrought in some pre-existent vegetative shoot.

So long as we direct our attention solely to the Seed-bearing (or Flowering) plants, and are prepared to pass lightly over all comparisons with Vascular plants lower in the scale, this opinion may stand. But as the 19th century drew on, the knowledge of the lower forms was greatly widened, especially in the case of such plants as the Ferns and Club-Mosses. This supplied the material necessary for a revised theory of the origin of the flower. Moreover, about the middle of the 19th century, views of descent began to make their influence felt upon the study of form in plants. But it was only slowly that this became apparent in a fuller knowledge of the genesis of the flower of the higher plants. Consequently it has only been in recent decades that the acceptance of the flower as the result of metamorphosis of a vegetative shoot has ceased to satisfy. It will be well to explain the reasons for the change.

In a previous chapter (p. 44) it has been shown that the life-story of a Fern consists of two phases, the relatively small green prothallus and the relatively large spore-bearing Fern plant. It was Hofmeister who first recognised that a similar alternation of two such phases is characteristic of all Vascular plants, including all those which bear flowers. In all of them that obvious thing which we call "the Plant" is the spore-bearing generation, and its end is the formation of sporangia and spores. What we are now considering is the evolutionary relation of the sterile and fertile regions of such plants.

In every normally completed life of Vascular plants sporangia are formed. This is so in all living Ferns, Club-Mosses, and Horsetails, as well as in Seed-bearing plants. Moreover the evidence of the fossils shows that it was the same with the related plants of the distant past. Hence we may conclude that it has been so throughout their descent. If this be true, then the origin of the Flower, and particularly of those sporangial organs which are its distinctive feature, cannot have been an afterthought. They cannot ever have been superposed upon a pre-existent vegetative system. And this must still be true notwithstanding that in the life of any ordinary plant the vegetative leaves appear first. The Flower cannot have been a mere result of metamorphosis of a vegetative shoot, for it represents a phase which has grown up together with the rest of the plant, and has been an integral part of its completed life through-out descent. Those who realised this to the full were at first disposed to invert the theory as stated by Goethe and his followers. Instead of holding that the flower was really a vegetative shoot adapted for propagative duties, they were disposed to assume that the whole plant had originally been propagative, and that the vegetative system had originated by sterili-sation of certain of its parts. It was suggested that such sterile parts were then told off to purposes of nutrition as a new duty. Moreover, experimental evidence from the Ferns seemed to tend directly towards this conclusion, for it was found possible experimentally to transform fertile leaves of Ferns into sterile leaves, with abortion of the sporangia. This seemed essentially parallel with the transforma-tion of stamens into petals, or carpels into green leaves, as is seen in some " doubled " flowers.

It is by no means outside the bounds of possibility
that this theory of sterilisation may be near to the
ultimate truth. But at present we are not in a
position to look with any certainty so far back as to
the ultimate origin of either the vegetative or the
propagative system of the spore-bearing plant. A
third view which does not raise questions of ultimate
origin has, however, been suggested by a comparison
of the simpler Vascular plants. It appears to stand
upon a sound basis of fact and of logical reasoning,
for it starts not from any hypothetical forms, but
from such plants as we can see living to day. It is
that the whole shoot of relatively primitive Vascular
plants was non-specialised : it served as a general-
purposes shoot. Both functions had to be performed
by it in earlier times. But in the course of evolution
of higher types differentiation took place, so that a
certain region became specially developed to carry
on nutrition, another region of the shoot produced
spores. The result of this is that the two functions
are carried out by distinct regions of the shoot
specially developed for the purpose. The vegetative
region habitually comes first in the individual life,
the propagative region, or " flower " as it is called in
the higher plants, appears only late after the plant
has acquired sufficient material for the formation
of the propagative organs. It is by comparison
of Seed-plants with the lower Vascular plants that

Fig. 11. Leaf of the male Shield-Fern (*Nephrodium filix-mas*, Rich.), about one-eighth natural size, the lower part having the under surface exposed. To the left a single segment, bearing the sori, each covered by the kidney-shaped indusium. (After Luerssen.)

evidence may be obtained of the correctness of this view.

Two common plants, which are widespread in Great Britain, may be cited as illustrative examples, viz. the Common Shield-Fern (*Nephrodium filix-mas*), and a common Club-Moss (*Lycopodium Selago*). In the former the plant consists of a simple shoot bearing leaves which are all alike, but large and freely branched. Though in young plants the leaves bear no sporangia, in any well nourished plant that is mature all the leaves are as a rule fertile. They are in fact leaves that serve various purposes, for when young they protect the tender apical bud, and when expanded they serve the double purpose of nutrition and of propagation by spores (Fig. 11). But in Flowering plants these three functions are generally carried out by three different types of leaves, the scale-leaves protecting the bud, the foliage leaves serving for nutrition, and the stamens and carpels for propagation. In some Ferns these distinctions are already suggested; for instance, in the common Hard Fern (*Lomaria spicant*) the sterile and fertile leaves differ in outline though fundamentally alike: and in the Royal Fern (*Osmunda regalis*) there are some stunted leaves which serve for protection only. Such occasional distinctions seem to suggest that our hypothesis of differentiation is really a correct one.

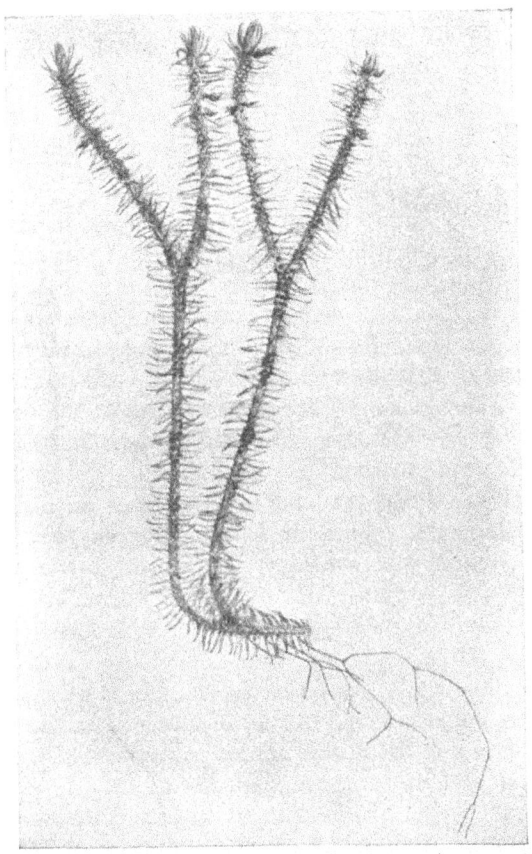

Fig. 12. Plant of *Lycopodium Selago*, showing on its several axes
alternating sterile and fertile zones. This is best seen about
half-way up the specimens. The leaves are alike in both; the
difference lies in the absence of the sporangia from the sterile
zones. (Slightly reduced.)

In the Common Club-Moss (*Lycopodium Selago*), after an initial vegetative growth in which all the leaves of the young plant are sterile, the shoot settles down to a condition which is shown in Fig. 12. The small and simple leaves appear all alike in size and texture, being clearly nutritive in function. The whole shoot is uniform throughout its length, except that it is marked off into successive zones which are alternately sterile and fertile. In the former a single sporangium is present in the axil of each leaf : in the latter the sporangia are absent. But at the limits of the fertile zones there are smaller abortive sporangia, while occasionally an isolated sporangium, or two or three of them, may be found within an otherwise sterile zone. When it is remembered that spore-production was constant throughout descent, such facts point very directly to the conclusion that there existed at first a shoot which was both nutritive and propagative, and that this Club-Moss shows those functions only partially separated, the sporangia being imperfect or entirely suppressed in the sterile zones, while the fertile zones maintain their pristine state. But in the great majority of the Club-Mosses the sporangia are associated with the leaves at the end of the shoot only, forming a cone or strobilus. It is plain that in them the differentiation has been carried out more perfectly. Such cones are the prototypes of Flowers, while the sterile region below represents the foliage as seen in the higher Seed-plants. So far then from the Flower being the result of "Metamorphosis" of a vegetative shoot, as the theory of Goethe maintained, it appears more probable that both arose from a common source. If any transformation or "metamorphosis" has occurred, it was probably a conversion of leaves which were primitively fertile into a sterile state, by abortion of the sporangia which they bore. And we have seen that this change can be brought about experimentally even at the present day. But that is the direct opposite of what was designated by Goethe as "progressive metamorphosis." It may perhaps be called with more propriety "Sterilisation," a term which indicates clearly what has probably been a prevalent phenomenon in the course of descent of Vascular plants.

Quite recently a direct support for such opinions has been supplied by the discovery and description of the flowers of the fossil Bennettitales. These, as also the modern Cycads to which they were allied, were linked by many characters to the Ferns: but on the other hand they produced seeds resembling in certain essential points those of the modern Pines, and Larches, etc. These Cretaceous and Jurassic plants showed a decidedly Fern-like character of the staminal leaves, while the centre of the flower was occupied by seed-bearing structures forming collectively a sort of pistil (Fig. 13). Such flowers were borne on the ends of lateral branches, and terminated their growth just as those of Seed-plants do nowadays. The foliage resembled that of the present-day Cycads. The most interesting point in these flowers as bearing on our present discussion is the similarity of their stamens to the fertile leaves of certain Ferns. The view which may be taken of these primitive flowers is that they represent a region of the Fern-like shoot which has retained its function of propagation, while the region below is sterile and is developed to carry on the other pristine function of nutrition.

Such an example as this indicates the origin of the propagative organs at the centre of the flower. But so far no explanation has been given of the origin of the outer envelopes,—the calyx and corolla : nor is there any certain knowledge how they came into being. It may be that they originated in different ways in the various groups of plants. In some cases it seems probable that the calyx sprang from bract-like leaves of the vegetative region already sterile, which became specialised for a protective duty. On the other hand, the corolla in certain cases was probably derived from the outermost stamens which became diverted from their specific function as propagative organs to the duty of attraction of insects by colour, scent, and honey-secretion. By such means a further step of sterilisation appears to have been

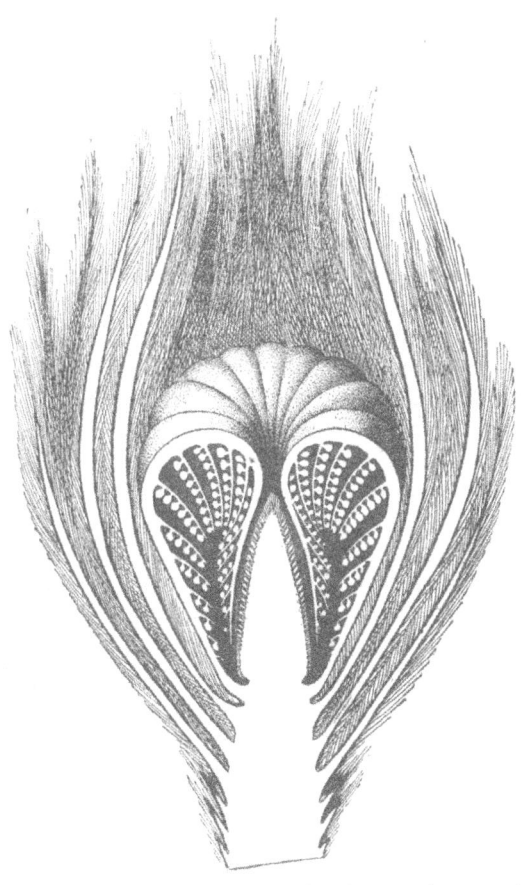

Fig. 13. Restoration of an unexpanded bisporangiate strobilus, or flower, of *Cycadeoidea*, with part of the enveloping hairy bracts removed. The unexpanded fronds of the staminate disc are curved inwards, and show the reduced fertile pinnae. At the centre is the apical cone covered by short-stalked ovules and inter-seminal scales. About natural size. (After Wieland.)

effected subsequent to the earlier segregation of the nutritive from the propagative regions of the shoot.

Thus we may figure to ourselves how the flower as distinct from the vegetative region originated in the higher plants. The main point to bear in mind is that the propagative function recurred in each fully completed life-cycle throughout descent. It was never an innovation. The propagative organs cannot at any time in descent have been superimposed upon a pre-existent vegetative system. Their appearance may have been deferred in the individual life, and it is only natural that the nutritive phase should precede the propagative, since food-material is necessary for the latter process. But however long the nutritive phase might be, or however prominent the organs which carry it out, the propagative phase supervenes sooner or later in every completed life. All evidence points to the belief that it was represented throughout the course of evolution.

Once the two pristine functions were allocated to distinct regions of the shoot, these were open each to its own distinct specialisation, and thus they became fitted for their respective duties. And so it comes about that while in simple cases there may be some similarity between the flower and the foliage-shoot, as for instance in *Magnolia*, which is on that account held to be a relatively primitive type of Flowering plant, the two may diverge widely in their characters in more advanced types, as do the foliage and flowers in such plants as the Orchidaceae. But however greatly they may differ, we must still hold the flower and the vegetative shoot to be the ultimate consequences of segregation of parts of the same original type of shoot, to serve the two fundamental and constant duties respectively of propagation and nutrition. This is the most natural interpretation of those structural resemblances, which are so obvious, between the foliage-shoot and the flower which it ultimately bears. There is no need now for any theory of metamorphosis to explain the origin of the Flower. As first stated by Goethe it was a theory of mysticism, which was doomed to dissolution so soon as the disclosure of the necessary facts made comparative study possible.

CHAPTER V

POLLINATION AND FERTILISATION

By many of those who write for the general public Pollination and Fertilisation are used as synonymous terms. This may have been natural when, over a century ago(1793), Sprengel published his novel observations under the title of *Das entdeckte Geheimniss der Natur im Bau und in der Befruchtung der Blumen* (The Secret of Nature discovered in the Structure and Fertilisation of Flowers). But at the present day there is little excuse for such laxity. Even Müller's classical work appears in its English edition under the title of *The Fertilisation of Flowers by Insects*. And yet on opening it we find that it deals throughout with the transfer of pollen, and the actual detail of the act of fertilisation is not mentioned. It is well to be clear at the outset as to the correct use of these words when applied to the Higher Flowering plants. By *pollination* is meant merely the transfer of pollen from the anther, where it is produced, to the receptive surface of the stigma (Fig. 15, p. 73), where it is to develope further. By *fertilisation* is meant the actual coalescence of two cells, the one derived from the development of the pollen-grain, that is the *male cell*, the other contained within the ovule or future seed, that is the *female cell*. It will be at once apparent that some interval of time must elapse between the two events of pollination and fertilisation, an interval which is usually short, but may in extreme cases be as long as a year, or even more. Pollination in the Flowering plant precedes fertilisation, and is a means to that end, but fertilisation is the end itself.

It is not the object here to detail the various methods by which pollination is secured. The chief facts are already available in many published works, some of them written for scientific, others for popular readers. In all of them the importance of intercrossing is brought forward, involving the transfer of pollen from one individual to another, a process which brings with it definite racial advantages. The specialisation of the fertile region of the shoot for this purpose is as varied as it is often beautiful. The very genesis of the forms of flowers, their tints, and scents is in strict accordance with their efficiency as pollinating mechanisms. Our own aesthetic pleasure in them is purely subjective. It must not be permitted to blind us to this simple fact. At the same time we see that the expenditure of material in the production

of flowers is great. This should impress weightily
upon the mind the importance which attaches to
intercrossing.

Due recognition is readily given in popular books
to the wonderful mechanisms by which intercrossing
is carried out.   But it often happens that the actual
process of fertilisation, as it is effected in the Higher
Flowering plants, is left entirely out of account.   Its
main features are, however, of the highest importance
as giving material for comparison leading to broad
views of descent.   It will therefore be necessary to
recount how fertilisation is carried out in an ordinary
Flower.   A basis of comparison will thus be provided
with plants lower in the scale.

The stamen is an organ which usually bears four
pollen-sacs.   Each contains many pollen-grains, which
are set free by rupture of the protecting wall.   The
grains are commonly dry and dusty, but incapable of
spontaneous movement : so that each is independently
open to transfer by external agencies.   The numbers
of the grains are roughly in relation to the precision
of the removing agent.   Where the method is haphazard, as in the case of transfer by the wind, the
production of pollen is usually profuse.   Where the
method is exact, as in the case of insect-agency
working in relation to a highly specialised flower,
the production of pollen is limited.   Extreme cases
of economy are seen in Orchis where a single stamen

suffices for the requirements, or *Salvia* (Sage), where two half-stamens achieve the end. In either case the flower-mechanism for making use of the visits of certain insects is a peculiarly perfect one.

The pollen-grain itself is protected externally by a wall often thick, variously sculptured, and coloured.

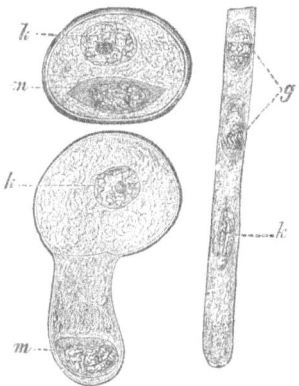

Fig. 14. Pollen-grain of *Lilium Martagon*, and its germination, $k =$ nucleus of the large vegetative cell of the pollen-tube: $m =$ small mother-cell of the antheridium: $g =$ generative cells, or male gametes. (× 375.) (From Strasburger, after Guignard.)

Within are found at the period of ripeness two cells, each with a nucleus (Fig. 14). The larger cell is styled the vegetative cell, and it takes no direct part in fertilisation. The smaller is called the antheridial-mother-cell. It consists of rather dense protoplasm

surrounding a nucleus. It is from this cell that the
gametes or male cells are derived.

The gynoecium or pistil occupies the centre of
the flower. It consists in a simple case of the parts
shown in Fig. 15. The lower enlarged region is
the ovary, the wall of which (*fw*) protects a single
straight ovule. The ovary is continued upwards into
a narrower region of the style (*g*), and is enlarged at
the tip into the receptive stigma (*n*). It is here
that the pollen is received, and retained upon the
roughened and often sticky surface. The ovule thus
enclosed in the cavity of the ovary consists of a
short stalk (*fu*), bearing the body of the ovule.
From the upper limit of the stalk arise two over-
lapping integuments (*ie, ii*), which closely invest the
massive nucellus (*nu*). This is an oval body of tissue
enveloping the large embryo-sac (*e*), which has
complicated contents. The most important of them
is the ovum, or egg, which is the largest of three
cells forming a group called the egg-apparatus (*ei*),
which is fixed at the uppermost end of the embryo-sac.
The ovum is a primordial cell, that is, it has no
cell-wall, and consists of cytoplasm with a nucleus.
It is the female cell, and it developes no further unless
fertilised.

The sexual cells or gametes are thus produced
apart from one another. The problem is to bring
them together when neither is motile, and one is

deeply enveloped in protective coverings. The usual
isolation of the pollen-grain allows of its free transfer,
and the varied mechanisms of flowers secure that the

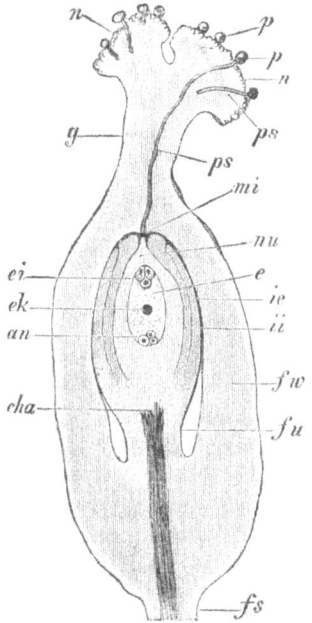

Fig. 15.   Ovary of *Polygonum convolvulus* during fertilisation.   *fw* =
wall of ovary : *fs* = stalk-like base of ovary : *fu* = funiculus : *cha* =
chalaza :   *nu* = nucellus :   *mi* = micropyle :   *ii* = inner,   *ie* = outer
integument : *e* = embryo-sac : *ek* = central nucleus of embryo-sac :
ei = egg-apparatus : *an* = antipodal cells : *g* = style : *n* = stigma : *p* =
pollen-grains : *ps* = pollen-tubes.   (× 48.)   (After Strasburger.)

pollen shall be landed upon the surface of the stigma (Fig. 15, p. 73). Thus the more hazardous part of the journey is accomplished, with or without intercrossing. Then begins the further development of the pollen-grain. A delicate tube grows out from it, which penetrates the tissue of the style, and makes its way downwards to the cavity of the ovary. Finally, it reaches the apex of the ovule, and comes into close contact with the egg-apparatus. This tube conveys the contents of the pollen-grain, so that the transfer to the deeply seated egg, otherwise difficult, is readily accomplished. When the germination of the pollen-grain upon the surface of the stigma begins, the contents of the grain pass into the tube. The antheridial cell divides into two gametes (g, Fig. 14), and these preceded by the vegetative nucleus (k, Fig. 14), maintain a position near to the apex of the tube as it grows. Finally, on reaching the embryo-sac, the nucleus of one of the gametes enters the ovum, and fuses with its nucleus. This is the essential point in fertilisation. The resulting cell, or zygote, forms the starting-point for development of the new individual. Since the nucleus is believed to be the bearer of hereditary qualities, the zygote will have received those of both parents, so far as they are conveyed by the nuclei which coalesce ; and those qualities will pass on to the newly produced offspring.

It is well to bear in mind that the event of fertilisation is probably the most critical in the whole life-story. There is no doubt that it figured in an equally critical way throughout evolution, and was a crisis in each completed life-cycle.

It is then right and fitting that a premier place should be given to the comparison of methods of fertilisation in discussions on descent. At first sight there may seem to be little relation between the propagative method of a Flowering plant here described and that of a Fern already detailed in Chapter III. So long as the external features only are noted the divergence appears to be a wide one. It is only when the details are followed out that an essential similarity appears. As a matter of fact an adequate knowledge of the propagative process in Ferns long preceded that of the corresponding events in the Flowering plants. It remained for modern methods of preparation to show how the nuclei behave, and it is upon this that the whole comparison turns. We now know that in the fertilisation of the Fern, just as much as in that of the Flowering plant, the essential point is the coalescence of two cells, and especially of the nuclei of two cells, which are more or less distinct in their origin. In the Fern the spermatozoid conveying the male nucleus makes its own way to the nucleated ovum. In the Flowering plant the male gamete is non-motile and is handed on passively by the pollen-tube to the ovum. The detail of method differs in the two cases, but the essentials are the same in both.

Having thus recognised their fundamental likeness, it lies open to us to enquire whether the differences afford any ground for opinion as to descent? The main difference between the two processes lies in the medium of transfer. In the Fern the spermatozoid moves in water, and external fluid water is necessary for fertilisation. In the Flowering plant it is not. The formation of the pollen-tube and its conveyance of the male gamete to the egg may be carried out without reference to external fluid water. It commonly occurs in dry summer weather, and is only dependent on water so far as the whole plant is so dependent for carrying out its ordinary functions. In this respect then the Flowering plant may be said to be essentially a plant of the Land. The Fern, though capable of vegetating on the Land, is dependent at a most critical moment on external water. In this it reveals an amphibious character.

But a further question which has its immediate bearing on descent is whether there is evidence of transition from one of these types to the other. As the Ferns show many relatively primitive characters both in structure and development this question will present itself only in one way. Is there any evidence of a transition from the aquatic fertilisation by spermatozoids to the terrestrial method by a pollen-tube? A certain degree of probability that such a change had taken place was derived long ago from the general comparison of the propagative methods in Ferns and Flowering plants. But the demonstration of spermatozoids still surviving in some

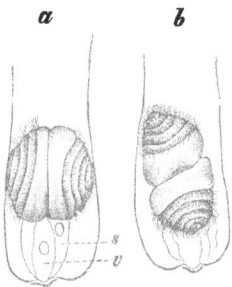

Fig. 16. *Zamia floridana.* Distal end of the pollen-tube, showing the vegetative prothallial cell (*v*), the sterile sister-cell (*s*) and the two spermatozoids. *a*, before movement of the spermatozoids has commenced; *b*, after the beginning of ciliary motion: the prothallial cell is broken down, and the separation of the two spermatozoids is taking place. (× circa 75.) (From Strasburger, after H. J. Webber.)

Seed-bearing plant would naturally appear more conclusive than such comparisons could possibly be. Some botanists had long been expectant that instances of this would eventually be found. The expectation has now been realised in the case of two ancient families of plants belonging to the Gymnosperms. Such plants, as their name implies, show their primitive character by bearing their seeds exposed, as in the Pine or Yew, and not contained in a seed-vessel, like the more advanced Flowering plants. It is in the Cycads and in the Maiden-hair Tree (*Ginkgo*) that fertilisation by motile spermatozoids has been found to occur. In both cases the pollen-grain germinates essentially as in other Flowering plants. But the gametes which each produces, instead of being passive, are motile, by means of cilia, in fluid extruded into a hollow chamber at the apex of the ovule: and thus access is gained to the egg (Figs. 16, 17). The conclusion which follows from such facts is unavoidable. The condition of the Cycads and *Ginkgo* can only be held to be vestigial; moreover the motility of their gametes is of little practical use. The gametes of the higher Flowering plants must then be regarded as spermatozoids which have given up this unnecessary motility, and become passive. Thereby they have parted with the last indication of their pristine mode of fertilisation by means of gametes freely swimming in water, and show full accommodation to life upon Land. But the Cycads and *Ginkgo* still show traces of their amphibial origin. The rest of the Seed-plants, however, now constitute a true Flora of the Land, characterised as such by their complete independence of external fluid water as a medium for fertilisation.

The importance of the step thus taken is very great. The most convincing indication of it is seen in the numerical richness of the types which have been emancipated from the restrictions imposed upon Land-growing plants by fertilisation through the medium of water. They exceed all other Vascular plants both in number of species and of individuals,

Fig. 17. *Zamia floridana.* Mature free-swimming spermatozoid, or male gamete. (× 150.) (From Strasburger, after H. J. Webber.)

and form the dominating Flora of the present day. On the other hand, the Cycads and *Ginkgo* show all the signs of being archaic types. For the Cycads are now represented by few genera, and are almost confined to the Tropics, though formerly more widely distributed; while *Ginkgo*, which in earlier geological periods was a relatively prevalent type, exists now only as *G. biloba*, the Maiden-hair Tree of China and Japan. Such facts give additional weight to the conclusion which has been drawn. They show in the form of practical results how essential that emancipation was, in order that full advantage might be taken of the opportunities offered by life upon exposed land surfaces.

CHAPTER VI

FIXITY OF POSITION AS A FACTOR IN THE
EVOLUTION OF PLANTS

WHENEVER there is heavy weather on our coast
combined with unusually high tides the surf disturbs
the sand collected in the rocky inlets, and as likely as
not lays bare the roots of land-plants which straggle
down to the high-water level. Here as elsewhere the
vegetation of the land encroaches upon the shore
with some degree of regularity of zonation, so that it
may almost reach downwards during the quieter
seasons to the usual high-tide mark. After such a
favourable period, if a heavy sea comes breaking in
at the full of a spring tide the encroachment is apt to
be checked, the foremost of the advanced guard are
uprooted, and the roots of those that pressed behind
are exposed. A glance at the wreckage demonstrates
at once the extent and the intimacy of the attachment of the ordinary land-vegetation to the soil.
It impresses more forcibly than any contemplation of

plants seen in the growing state the fixity of position which is so prevalent a factor in plant-life.

It was long ago laid down as a general distinction between animals and plants, that plants live, while animals live and move. This antithesis still survives in the mouth of the artist when he includes plants under his designation of "still life." It is hardly necessary to-day to refute this distinction, or to show how far it is from accordance with fact. Every elementary student knows that active life and movement in some degree are inseparable. The movement may be molecular only, or intracellular: in either case it would be invisible to the naked eye, but still it would be there. Moreover, the student soon learns how closely such movements depend for their activity upon those conditions which favour vigorous physiological change : for him movement of some sort is a common expression of life in plants as well as in animals. Darwin's well-known book on the Power of Movement in Plants has gone far to disillusion the public on the point. He there showed that, apart from the more rapid and obvious movements in such cases as the sensitive plant (*Mimosa*), ordinary seedlings and growing plants at large execute movements. These, being slow, are for the most part indistinguishable by the naked eye, but they may be readily demonstrated by simple means. Thus we approach the conclusion that even plants move in greater or less degree during active life, and that "still life," as applied to them, is in point of fact a contradiction in terms.

Nevertheless there is an underlying truth in the old distinction between animals and plants, though it applies only in a rough sense to the higher terms of the two series. Fixity of position is the common lot of the higher plants just as an ambulatory habit is the rule for the higher animals. I propose, on the one hand, to examine the circumstances which may have led to the adoption of the fixed habit in plants, and on the other to point out certain disabilities which it imposes upon them. These have had to be overcome before success could be assured among the thronged life which subsists on the earth's surface.

It is through the study of nutrition and mechanical construction that the circumstances are to be sought which have led to the adoption of the fixed habit by plants. The two kingdoms differ widely in their methods; for while animals take their organic supply at second hand, absorbing in some form or another material already elaborated from its inorganic sources, capturing it, and ingesting it often in bulk ; plants go direct to those inorganic sources, and elaborate their food themselves, absorbing its materials always molecule by molecule. One consequence of this fundamental difference in the method of nutrition is that

in the plant the complete enclosure or encystment of
the primitively naked protoplast by a cell-wall presents no serious obstacle, provided that the cell-wall
be permeable to the molecules of the food to be taken
in, these being in the condition of solution in water.
But encystment brings great mechanical advantages,
and is indeed the chief means by which the construction of the large tissue-masses of the plant becomes
mechanically possible. So long as the wall is
sufficiently permeable, there is no need for any part
of the protoplast to be freely exposed, and as a
matter of fact, except among the simplest organisms,
or the reproductive cells of those which are more
complex, the cells of the plant-body are habitually
encysted (Fig. 18). But the cell-wall which is thus
a general feature in all but the simplest plants
presents at once an obstacle to free movement. The
protoplast is fettered by its armour more seriously
than any mediaeval knight. It has sold its birth-
right of free external movement for the pottage
of mechanical protection and structural stability.
Among the simpler aquatic organisms many naked
plant-forms are found, which like the simpler animals
are capable of active movement. But there are
others in which the movement is only an occasional
incident connected with propagation, others again in
which the encysted state is permanent. These may
be held to illustrate the sort of steps which have led

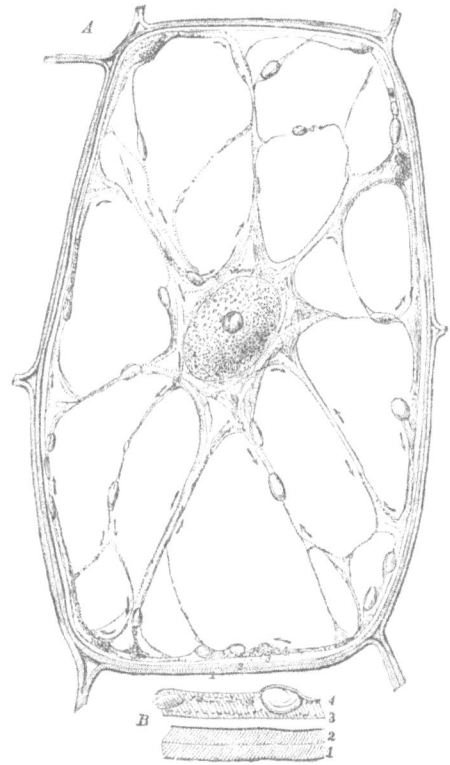

Fig. 18. *A*, parenchymatous cell, highly magnified : at the exterior
is the cell-wall (shaded) which encysts the body : the latter is
composed of a protoplasmic lining which surrounds the large
watery vacuole, while in the centre the nucleus is suspended by
bridles of protoplasm. The arrows indicate the movements of
the protoplasm during life. *B* shows a portion of the cell-wall
and protoplasmic lining more highly magnified. (From Vines'
Lectures on Physiology of Plants, after Hanstein.)

from the primitive freedom of the protoplast to the bondage of encystment.

The protoplast thus encysted remains isolated only in the simplest forms. In all the more complex the plant-body is made up of many, it may be of millions of individual cells, derived ultimately by division from a common source. Each is united with its neighbours so as to constitute a tissue, or tissue-system. The bondage of each encysted cell applies collectively to the whole mass of them, and a large, relatively immobile plant-body is the result. It is no wonder that with tissues thus constituted the movements of plants even in the young state are slow, and act within narrow limits : and that as the tissues become mature and their cell-walls thicken, the parts composed of them become rigid, and lose the power of movement altogether.

The nutrition of such encysted cells or cell-masses, in the case of the Algae which inhabit fresh or salt water, may take place at any point of their surface which is exposed to the surrounding medium. Such Algae are in simple cases suspended without attach-ment, and constitute the large proportion of the so-called "plankton," or floating life. They may be collected in quantity from the surface of salt or fresh water, in tow-nets of muslin drawn in the wake of a boat. The collection will be found to consist for the most part of relatively simple organisms of small size.

But some few plants of larger growth, such as the well-known Gulf-Weed (*Sargassum*), may also float unattached, growing and taking nourishment at the surface of the open sea. This is, however, exceptional, and for the most part the more advanced vegetation of salt and fresh water is attached to the substratum, though the reason for this is not directly one of nutrition, for they still feed by absorption at the general exposed surface. The fact that in the sea, in large bodies of fresh water, and in streams the medium in which the plant lives is always changing owing to the tides, currents, and general water-movements, materially aids nutrition, for by this means fresh supplies of inorganic materials are continually being presented to the floating plants.

Coming finally to the vegetation of the land, we find it composed essentially of plants rooted in the ground. The extent and effectiveness of the attachment are well seen when the roots are washed clear of the soil, as on the shore about high-tide mark, or along the banks of a stream that has been in spate. The reasons for this fixity by the root are primarily mechanical and nutritive. It is by such anchoring of the plant firmly in the soil that it is possible for it to raise its limbs and leaves upwards to expose them to light and air. And, as every one knows, this is a condition necessary for the successful nutrition of the green plant. But, secondly, it is in the root-system that the absorption of water and of the salts dissolved in it is localised.   These are abstracted from the soil by the roots, to be carried by the transpiration stream to the foliage, where the water evaporates away, leaving the salts in the very region where the construction of new organic stuffs is proceeding.   Moreover the supply of food stuffs is further facilitated by constant movement of water in the soil.   This, together with the growth and spreading of the roots, and the recurrent renewal of their absorbing surfaces, result in conditions which bring the plant in touch with ever fresh supplies of raw materials.   It is thus seen that the fixed position of the plant in the soil is a necessary condition of successful nutrition in all the ordinary vegetation of the land.   A measure of its effectiveness may be seen in the permanence of large trees, such as the *Sequoias*, or Mammoth Trees of California, which bear evidence of individual growth for over 1000 years.

The facts alluded to in the preceding paragraphs are so familiar that one is apt to neglect the important restrictions which necessarily follow from them.   In considering these it will be well to compare the higher plants, which are typically fixed to the substratum, with the higher animals which are ambulatory in the habit.   So we shall see actually worked out in practice the consequences which this difference in habit has entailed.   The most important disabilities under which the stationary plant suffers as compared with the freely mobile animal are four: (1) difficulties of self-protection against animal attack, and (2) against climatic conditions : (3) difficulties in effecting pollination and intercrossing, and (4) in the distribution of the new germs when produced. These will severally be considered, and the methods will be explained by which plants have met the problems raised by their immobility.

The stationary plant is the natural victim of the mobile herbivorous animal, and in many cases, such as the ordinary pasture grasses, clover, or plantain, there are no special means of defence. The caterpillar, the slug, and in tropical countries the leaf-cutting ant all take their toll, as well as the larger grazing animals, and frequently without any reprisals. The only resource of the victim is to restore by renewed growth as quickly as possible the parts that have been lost. The animal, on the other hand, which is the victim of a predatory attack has the chance of escape by a combination of wariness and speed, but this is denied to the fixed plant. Other means, however, of protection of the plant against predatory animals exist, and are effective. Some plants, such as the Spurges and Asclepiads, contain distasteful substances in their milky juice : others, such as the Dock and Wood-Sorrel, and some Begonias are protected by their acid taste : others such as the

Rue and the Sweet Rush (*Acorus*) by ethereal oils :
others again are protected by sharp needle-shaped
crystals embedded in their tissues.   These, which are
especially effective against the attacks of snails, are
found in many Monocotyledons, such as Narcissus or
Orchis, and in some Dicotyledons, such as the Vine.
The sharp crystals are usually associated with mucil-
age, which swells on access of water when the tissues
are crushed or cut by chewing or biting. The crystals
are thus dislodged.   Pointing irregularly in all directions, and puncturing the delicate tissues of the
mouth, they make the plant distasteful, and thus
protect it from further attack.   The experiments of
Professor Stahl with snails have shown how effective
this protection actually is, even when the snails had
been famished for the purpose of the test.   In other
cases the protection may be by pointed hairs, spines
or thorns, and this is particularly prevalent in dry
climates, where the succulent character of the vegetation would offer peculiar attractions to animals.   The
most conspicuous examples are among the Cactaceae.

These notes must suffice to suggest some of the
many ways in which plants, owing to relatively minor
characters which they possess, are enabled to subsist
even where they are exposed to the attacks of
herbivorous animals.

Similarly they have to meet the risks of climate.
The mobile animal may protect itself against these

risks in various ways : such as by migration, or by
retiring into nests or burrows. The fixed plant has
to remain where it is. It may hibernate, however,
underground, as do many herbaceous and bulbous
plants; or it may reduce its physiological risks by the
fall of the leaf, as do our ordinary deciduous trees in
the autumn ; or it may elude the period of extreme
conditions in the form of seeds or resting spores,
which are shown to be more resistant than the
actively vegetating plant. But it would take too
long to do more than merely suggest these as some
of the many devices by which plants overcome the
difficulties of climatic stress, to which their fixed
position condemns them.

A still more weighty disability than those which we
have thus seen to follow on the adoption of a fixed
position, is the difficulty in effecting pollination, and
its usual concomitant intercrossing. This is not the
place to discuss the advantage which follows on cross-
fertilisation in plants. It may be accepted as a general
principle that advantage does accrue. A general
reference may be given to Darwin's book on *The
Effects of Cross and Self-fertilisation in the Vegetable
Kingdom.* He there states, in his last chapter, that
"the first and most important of the conclusions
which may be drawn from the observations given
in this volume is that cross-fertilisation is generally
beneficial, and self-fertilisation injurious." In am-

bulatory organisms, such as the higher animals, the
search for a mate is readily carried out.   But where
the organism is fixed, that fact at once places fertilisation and still more intercrossing at a disadvantage.
It is obvious that in them special means will have to
be provided in order to ensure its taking place.
When the importance of intercrossing is fully realised,
as well as the embargo on its ready occurrence which
follows from the fixity of position, it becomes intelligible that the sacrifices to secure it should be
great, and the means employed various and even
recondite.

Few subjects within the scope of botanical science
have raised more interest for the general reading
public than the cross-pollination of Flowering plants.
Ever since the time of Sprengel, who first demonstrated
the importance of animal agency for this end, and
drew attention to the specialisation of flowers as
mechanisms which should co-operate with the active
insect, the interest has been on the increase.   But it
was specially stimulated in later years by the writings
of Darwin, and of Müller, and has been expanded by
the observations of many other authors.   It is now
a matter of common knowledge that not only are
insects, snails, birds and even bats made use of as
the carriers of pollen from flower to flower, but the
natural motor impulses of winds and moving water
are also employed.   It has also been pointed out how

mutually adaptable, and indeed mutually dependent, certain plants and insects are upon one another. No better example of this can be quoted than that of the genus *Aconitum* (Monkshood), and the humble-bees that pollinate it. The northern limits of distribution of the two in Asia coincide with an exactitude that can only mean an intimate biological relation. This is not the place to describe such observations in detail, but rather to consider the conditions which have made it necessary to resort to such artifices, and the evolutionary steps which may have led to their adoption. These conditions, simple though they may be, are frequently overlooked by those who pursue this branch of observation, or omitted from popular statements on the subject. Their recognition, however, must underly any rational study of pollination.

The Flowering plants of the land were without doubt evolved from Fern-like plants, and these again are generally held to have originated from the primitive Algal Flora of the water. In such organisms as these the male cells, and sometimes the female also, are capable of independent movement in water, into which they are extruded at maturity (compare Fig. 2). And if they come, as they frequently of necessity do, from different parent plants, the crossing is easily effected as a consequence of their own independent powers of movement. In the Pteridophytes such as the Ferns, we see organisms which, originating from an Algal source, have adapted themselves to life upon the land. But their sexual propagation is still through the medium of external fluid water, to which they have occasional access in case of rain or copious dew (compare Figs. 8 and 9). Their spermatozoids are capable of active transit through water, and their course has been shown not to be entirely at haphazard, but a chemically directive influence is exerted by the diffusion of soluble substance from the neighbourhood of the egg-cell, which attracts the motile sperm. In such cases the obstacles to intercrossing are not such as to require extraneous aid, and there are analogies with what occurs among ambulatory animals.

But it is otherwise with those organisms which have taken the further step in specialisation to life on the land involved in the adoption of the Seed-habit. In these a condition for fertilisation to take place is that by some means the pollen-grain shall be transferred bodily from the stamen where it is produced to the stigma, or in Gymnosperms to the ovule, or future seed itself. Since the stamen may be on a different branch, or even on a different plant from the ovule, a real transfer, often through a considerable space, is a necessary feature of propagation. The pollen-grain is itself incapable of spontaneous movement. And so some external agency must of necessity be called in if pollination is to be carried out at all.

The origin of Seed-plants having this requirement dates back to the most ancient epoch which has left to us any remains of terrestrial plants. The earliest Seed-plants (Gymnosperms) were probably as dependent for pollination upon external agencies as are the Seed-plants of the present time. Moreover, since in most of these the male and female flowers were distinct from one another, the carrying out of pollination was for them a condition of setting seed at all, not merely a condition of securing the advantages which follow from intercrossing. The call for external help was thus not merely a matter of advantage, but of imperious necessity.

There is no clear evidence how the earliest Seed-plants were pollinated. The fact that the pollen was transferred is proved by its observation actually within the orifice of the ovule, in sections of fossils of Carboniferous age. The grains were relatively large and smooth, so that though the agency of transfer may have been the wind, there was no definite specialisation to that end. Certainly the wind is the usual means of conveyance of the pollen in the Gymnosperms of the present day, though among them are some which rely upon insect agency; and it is quite possible that among their early correlatives the latter may have figured more largely than has hitherto been believed. But whatever may have been the case for the more primitive Seed-plants, it seems clear that in the Higher Flowering plants, animal agency was early available, and was soon made use of. There is ample evidence of insect life from the Secondary Rocks, and it was in that period that the outburst of evolution of Flowering plants took place. Though many of the simpler types employ the agency of the wind or of water, there is good reason to believe that the more complex flower structures were evolved correlative to and together with their insect visitors, a parallelism as probable as it is interesting. There may doubtless have been cases of transfer from the one agency to the other, as for instance in the Rue (*Thalictrum*), which has become wind-pollinated in a family (Ranunculaceae) as a rule insect-pollinated; and the converse is also possible. It is not our object here to define the actual degree of dependence for pollination upon one or the other of these two important agencies, but rather to indicate that, whatever the detail, this principle rules in Seed-plants :—that in very many cases to secure pollination at all, and in all cases to secure the advantages of intercrossing, some agency outside the plant itself is a necessity; and that this necessity arises primarily from the immobility imposed by the rooted habit.

The fixed position is then the prime factor leading to all those wonderful adaptations of the flower which have pollination as their end. It may also be held as a circumstance favouring, and so helping to explain, the prevalence of that type in flowering plants where stamens and carpels are present in the same flower. This juxtaposition makes self-pollination possible, and in many cases probable. There are indeed some flowers which do not open at all : in these self-pollination is virtually a certainty. Such arrangements may be held to be a provision against the risk of pollination failing. A higher degree of certainty of setting seed is thus achieved than would be possible by dependence upon extraneous agents. Thus taking a broad view of the matter it would appear that it is ultimately the fixed position which has made hermaphroditism more common among the higher terms of the Vegetable Kingdom than it is among the Higher Animals.

Following a similar line of thought we may enquire what is the reason for those varied methods by which plants secure the distribution of their seeds. The high importance of that distribution must first be grasped. In order that each germ may have the best chance of growing to maturity, it must on germination take up its position in the soil at such a distance from its fellows as not to compete with them for light or sustenance. In proportion as this is achieved with success there will be provision not only for the maintenance of the race, but also for its spread into regions hitherto unoccupied, with increased chance of survival as a consequence. In the case of the Higher Animals the necessary spread of the race is secured by individual movement, and accordingly they possess no special means for the distribution of their germs. A nest of young birds or rabbits disperses of its own initiative on approaching maturity. But in the case of the Higher Plants fixity is a condition of nourishment, and such distribution as is necessary must be secured for the germ while young. When once the germ is established as a rooted plant the opportunity is gone for ever. It need then be no surprise, however much the details may command our admiration, that the adaptive arrangements in plants for the distribution of their germs should be varied and effective. As in pollination, so also in seed-distribution the forces of nature, wind and water, are used : the former easily wafts away the light seeds of small size, such as those of the Orchidaceae, or Ericaceae, without special adaptive structure. But whole families, such as the Compositae, the Asclepiadaceae, and the Willows, and especially many trees such as the Sycamore and Elm, show definite mechanisms which secure distribution of seeds of larger dimensions, by flocculent hairs, or winged expansions, readily caught by the breeze. Water transit is proverbially cheap, and still larger seeds may be conveyed by floating away from the parent plant. This is seen in the Coco-nut, and as an extreme example, the double Coco-nut of the Seychelles (*Lodoicea*) may be cited, probably the largest single-seeded fruit in the Vegetable Kingdom.   Again, in some cases animal agency is used, externally by hooks and barbs on the surface of the seed or fruit, which catch upon the coat of a passing animal: or internally by pulpy developments, which attract the animal as food, and the seeds swallowed with the pulp are deposited with the excreta at a distance.   Or plants may develope mechanically effective means of propulsion of their own.   The common Broom and the hairy Bitter Cress have explosive fruits, which scatter the seeds far and wide.   An extreme case is that of the Euphorbiaceous Sand-Box tree (*Hura crepitans*), the woody fruit of which explodes with a report like a pistol shot, and scatters the relatively large seeds to a considerable distance.   Or the principle employed may be that of the common squirt, as in *Ecballium*, the Squirting Cucumber, which extrudes its semifluid pulp, seeds and all, to a distance of many feet, as it detaches itself from its stalk when ripe.   These are merely examples of the varied artifices employed by plants for the distribution of their germs.   The *raison d'être* of all of them is ultimately to be found in the fixity of position, which the plant assumes at once on germination.

It is thus seen that, in the first place, the encystment of the protoplast within a firm cell-wall, and secondly, the adoption of a fixed position by the plants of the ordinary land-vegetation for purposes of nutrition, have introduced certain disabilities into the life of the plant, and that these have had far reaching effect upon the course of further development. The early encystment, while it prevented the free movement of the whole body of those aquatic organisms which adopted it, affected their general biology in a comparatively minor degree. For it was still open to them to extrude individual protoplasts, for purposes of propagation, into the water in which they lived. And many of the lower forms do so habitually, thus securing by their motility both spread of the germs and opportunity for fertilisation. But with the spread of vegetation to the land, where the medium of water no longer gives free opportunity at all times for the movement of germs, the need for other means of spread became urgent. The results are seen in the multifarious methods of distribution of spores and seeds. The adoption of the seed-habit necessitated also some mechanical means of transfer of the pollen. This, together with the advantages derived from intercrossing, have led to the wonderful and varied devices for securing pollination. The prime cause of the difficulties was the encystment of the protoplasm, and the consequent immobility of the organism as a whole. Fixity of position has in fact dominated the whole course of evolution of plants.

The difficulties which beset pollination and dissemination have followed as secondary consequences. They have been overcome by those adaptations which command our interest and admiration. But too often the interest in the adaptations themselves has tended to obscure the prime cause which brought them into existence. It has been the object of this essay to place the cause and its consequences in their proper relation one to another.

CHAPTER VII

PLANT POPULATION

AN idle hour on a summer's afternoon may be pleasantly passed by any one who knows the Native Flora, in trying to identify all the plants within reach, as he sits, perchance, on some grassy slope. The vegetation of cliffs facing the sea, the turf on a mountain-side, the dense wiry growth of the golf-links, or the ordinary sward of a meadow would serve, all different as they are, to propound each its own conundrum. And happy may the observer feel who comes well out of an exhaustive study of all the plants great or small, complex or relatively simple, that grow within the radius of his arm's length. Often they will be in some stage of development which does not show distinctive characters, and this will itself complicate the problem, and provide tantalising uncertainties. But supposing the inventory to have been made with some degree of completeness, and all the various dwellers within the

limited circle to have been duly recognised, any
active mind will ask, Why are these plants here?
How did they get here? Why these instead of a
hundred others? You may perhaps be able, where
the ground is open and not under cultivation, to rule
out the direct hand of Man, so often a disturber of
the order of Nature. With some certainty it may
then be concluded that it is really Nature's own
achievement which has been observed and analysed.
How then does Nature bring it about that certain
species are present, and others absent from this
particular area of the earth's surface?

There are two main factors in this problem. It is
quite clear that as all living plants arise from their
like, in some way the germs of the plants recognised,
or of their ultimate parents, must have arrived within
the area observed. The first factor is then how this
area acquired the germs of the plants living there.
But further the growing plant developes from the
germ only when the conditions are favourable. From
the presence of any such growing plant it may then
be inferred that the conditions have been at least
tolerable. That they must necessarily be so is the
second factor in our problem. It is not proposed
here to consider the effect of external conditions
upon the growth and establishment of each germ as
an independent plant. That would involve some
review of the means of accommodation of plants to

their habitat : or perhaps rather the negative side of
the question might force itself more clearly upon the
mind, viz. the failure of germs to establish themselves
where the conditions are more severe than their
constitution will stand. This study would require to
be carried out with separate reference to each of the
species represented on the list observed, since the
reasons for the presence of each may have been
distinct and separate from those of others : indeed it
is probable that they were so. Such questions may
be allowed to stand aside for the present, and the
first factor will remain as the subject for discussion
here, viz. how the given area of ground acquired the
germs of the plants now established there. The question divides itself naturally again into two branches,
first the origination of the germs, and secondly, the
means of transfer of the germs themselves, or those
of their ultimate parents, to the spot where they have
developed into the plants which we see.

It will be obvious to any one who recognises fully
the ordinary forces of Nature, and especially the
movements of air and of water, and their secondary
effects upon the surface of the soil, and upon the
organisms growing upon it, that chance will enter
largely into the determination whether any given
germ, when separated from its parent plant and
exposed to their action, shall lodge finally in one spot
or another. The limited area within reach as we sit

may be taken as a unit, and considered among the
millions of like areas which go to form a country-side.
What is the probability of the light grain of some
grass, say of Holcus, Fescue, or of the Trembling
Grass, with its chaffy paleae serving as sails, being
carried by the wind, and finally coming to rest, till
under favourable circumstances it shall germinate?
Obviously the chances are very remote, but equally
clear is it that the larger the number of germs
produced the greater will be the probability. We
shall then enquire with interest into the fecundity of
plants, as estimated by the number of germs they
produce in a given time; for plainly this is a factor
which will go far in determining the plant-population as
we see it established in any area open to the free play
of natural conditions.

The third Chapter of Darwin's *Origin of Species*
dealt with the geometrical ratio of increase of living
things. Following Malthus he there pointed out that
"there is no exception to the rule that every organic
being naturally increases at so high a rate, that, if
not destroyed, the earth would soon be covered by
the progeny of a single pair." The effect of a
geometrical ratio of increase is apt to be under-
estimated; but a simple example makes it clear. If
a plant produces 10 seeds each year, and each of
these germinates, the result in six successive years
will be 10, 100, 1000, 10,000, 100,000, 1,000,000.

Plainly as the years go on the result becomes rapidly very large. Already Linnaeus had calculated that if an annual plant produced only two seeds, and their seedlings next year produced two, and so on, then in twenty years there would be a million plants. This is, however, a very slow rate of breeding, and it will be well to enquire what are the rates of production of germs seen in well-known plants. Taking first the case of Seed-plants, the following table gives the results of careful computation by Kerner of the number of seeds produced in a single season by an average specimen of each:

Henbane (*Hyoscyamus niger*)   10,000.
Radish (*Raphanus Raphanistrum*)   12,000.
Plantain (*Plantago major*)   14,000.
Shepherd's Purse (*Capsella Bursa-pastoris*) 64,000.
Fleabane (*Erigeron Canadense*)   120,000.
Tobacco (*Nicotiana Tabacum*)   360,000.
Flixweed (*Sisymbrium Sophia*)   730,000.

Extreme cases of productivity were found by Darwin among the Orchidaceae, and his estimates of the numbers of seeds were as follows:

	per capsule	per plant
Cephalanthera	6020	24,080.
Orchis maculata	6200	186,300.
Acropera	371,250	74,000,000.
Maxillaria	1,756,440.	

The Ferns also afford striking examples of an extreme power of productivity. I have estimated the number of spores of the Common Shield-Fern (*Nephrodium filix-mas*) produced by a well-grown plant in a single season at approximately 50 to 100 millions. But this is by no means the limit for plants of that affinity, as the subjoined table shows, in which the numbers quoted are estimated not for the whole plant, but for the single leaf:

Athyrium filix-foemina	16,000,000.
Polypodium aureum	96,000,000.
Cyathea dealbata	200,000,000.
Danaea	150,000,000.
Kaulfussia	173,000,000.
Marattia	2,800,000,000.
Angiopteris	4,000,000,000.

The productivity of Fungi is well known, and any one who has blown the spores off a fully matured culture of some common Mould, and seen the dense cloud that spreads into the air, would be prepared for high numbers on a rough estimate being made of them. In the common olive-green Mould (*Aspergillus herbariorum*), the number of conidia produced upon a single conidiophore or head is at a moderate estimate about 1500. These heads are formed in countless numbers in a few days upon a suitable nutritive medium. If 1000 of them were borne upon a single

square inch of its surface we arrive at the output of
1½ millions of conidia as the result of a few days'
growth over that limited area.

As a last example, taken this time from the
Bacteria, it has been stated by Cohn that in a suitable
medium at a temperature of 35° C. a cell of *Bacillus
subtilis* will take about 20 minutes to divide into two.
If this process be repeated continuously, and the
resulting cells all retain their vitality, the product of
a single germ would in a period of nine hours (*i.e.* in
a single night) amount to over 134 millions !

Kerner has illustrated the results which would
follow if such numerical increase as that we have
been contemplating were actually carried out unchecked, in the case of certain Seed-Plants. He
points out that if a Henbane plant developed 10,000
seeds in one year, and 10,000 plants sprang from
those seeds next year, and themselves produced
10,000 seeds each, by the end of five years ten
thousand billions of Henbane plants would have
come into existence. As the entire area of the dry
land of the earth is approximately one hundred and
thirty-six billion square metres, and there is room for
about 73 Henbane plants on one square metre, if all
the seeds referred to ripened and germinated, the
whole of the dry land would at the end of five years
be occupied. Or, in the case of the Flixweed
(*Sisymbrium Sophia*), the normal multiplication from

a single parent would if unchecked cover an area 2000 times as great as the surface of the dry land in the course of three years. A glance at the figures quoted for Orchids, Ferns, Moulds, and Bacteria shows that the Flixweed is by no means an extreme case of fecundity. It is hardly possible to express in words the consequences of successful establishment of every germ which such plants might produce within even a moderate period of time.

There is no need to astound or stupify the mind by the statement of further figures such as those above quoted. No brain can realise what is meant by a million units, and speaking of billions conveys no more than some general impression of vastness. What is desired is to make it clear that the production of potential germs is very great, and that even in cases of relatively low productivity the number of germs produced is far in excess of the actual requirement to make up directly for losses by death. There is in fact an immense overplus, which may be held as a very efficient reserve to meet, with an ample margin to spare, all the contingencies involved in the establishment of the germ, and its continued existence up to the time of its propagative maturity.

But there is another circumstance to be considered besides the mere number of the progeny. There is a diversity in different types of plants in the time required for the individual to reach the reproductive

stage. Clearly the shorter that time, other things
being equal, the greater will be the propagative power
of the species. The annual plant germinates, developes, fructifies, and dies all in a single season.
Sometimes in arctic or alpine regions (which, however,
are on that account little suited to the life of annuals)
that season may be only a few weeks in duration. On
the other hand, many plants live and grow for a long
term of years merely as vegetative individuals, enter-
ing late upon a propagative period. In some cases,
as the gigantic Talipot Palm (*Corypha*), or many Bam-
boos, the parent plant then dies. In other cases, as in
many forest trees, the fructification may be repeated
annually. In others again, as in the Common Beech,
a period of storage of nutriment may extend over a
series of years, to be terminated by a profuse fructification which suddenly depletes the store. Such
differences of behaviour modify, it is true, the details
of the propagative output, but the rate of increase is
as a rule so great that their effect is apt to be
swamped in the vastness of the totals, when taken
over an average of years.

In considering the chances of any given germ
becoming established as a new individual, the amount
of nutriment which it carries with it is a matter of
the highest moment. The larger the supply the more
thoroughly can the germ fit itself for acquiring its
own sustenance before the need for doing so actually

falls upon it. In the case of the spores of Mosses or
Ferns the store within the single-celled germ is small
indeed, but the grave risks which consequently attend
their germination are neutralised in great measure by
the enormous number of the chances of success.   In
this lottery the proportion of prizes to blanks is small,
but the issue of tickets is very large. In the seed-
bearing plants the store that accompanies each germ
varies greatly, from that of the minute seeds of an
Orchis or Rhododendron with their exiguous store,
to the huge supply carried in the Coco-nut. In the
former cases the seeds are very numerous : in the
latter a solitary large seed is matured from each
flower. In the case of large seeds the issue of lottery
tickets is not great, but there is a reasonable probability of each bringing a prize.

Obviously the mere production of germs is not all
that is necessary. If those of the same plant remain
in close juxtaposition, in cases of the large output of
germs, the young plants on germination would be so
crowded that an internecine struggle would ensue
between them. To have a good chance of success
each must be separated from its neighbours : dissemination is essential, and it will be effected with
the greatest difficulty where the seeds are largest.
The agencies by which this is carried out have already
been alluded to above (p. 98), viz. spontaneous
ejection, transfer by wind, and water, and by animal

movement. Putting on one side the spontaneous impulse of mechanical ejection as effective only within limited range, the most important agencies of transport are by wind, by water, and by animals. As a rule the consequences of their action are not very obvious owing to the ground being already tenanted by a mixed vegetation. It is only where a perfectly sterile area has to be peopled with a new covering of vegetation, or when a new organism gains access to an area previously untenanted by it, that the results can be fully appreciated. An experiment upon the grand scale was made in the formation of a completely new Flora of the Island of Krakatau, and fortunately its results were followed by competent observers who kept careful records. These form the best authentic story of the natural formation of a plant-population where none existed before.

Up to 1883 the islands forming the small group in the Sunda Strait between Java and Sumatra, of which Krakatau is the largest, were covered by dense vegetation. From May to August of that year successive volcanic eruptions resulted in the complete sterilisation of the surface, which was covered by hot stones and ashes. Thus on cooling an uninhabited desert was exposed, lying at a distance of fully 12 miles from the nearest living vegetation. In the intervening 25 years a new Flora has sprung up upon the islands. This has been studied at intervals: the

late Dr Treub, who visited Krakatau in 1886, concluded that the first colonists were blue-green Algae
associated with Diatoms and Bacteria. These formed
a suitable *nidus* for the spores of Mosses and Ferns,
and for the seeds of some Flowering plants adapted
for dispersal by winds. On the beach were found the
fruits and seeds of Flowering plants some of which
had germinated ; many of them belonged to the
characteristic strand-flora of the Malay Region. But
plants introduced by animals or by man were not
found by him on this visit, which took place only
three years after the eruption. In 1897 Penzig
visited the island, and estimated that of the Flowering plants noted $60.39\,^0/_0$ had reached it by ocean
currents, $32.07\,^0/_0$ by wind agency, and only $7.54\,^0/_0$ had
been transported by fruit-eating animals and man. On
a subsequent visit by a party of botanists in 1906, the
results as stated by Ernst show that though these
proportions for Flowering plants were not exactly
maintained, still the largest number were borne by
water transit, and the smallest by animal agency. If
the Ferns be added to the list of plants with wind-
borne germs, this would raise the estimate of the
Vascular plants which had arrived in this way to $37\,^0/_0$.
Thus for an oceanic island the most effective agency
of transit is water : wind-carriage takes a middle
place, and transit by animal agency is the least
effective of the three.

The results of successful dispersal combined with high fecundity in any given organism come out clearly in cases of invasion of areas previously untenanted by it. Darwin, in the *Voyage of the Beagle*, quotes the case of the Cardoon Thistle in various districts of South America. Its germs are spread by the wind. Though not a native, it has by this means occupied hundreds of square miles of country to the exclusion of the native Flora. Similarly in Ceylon, *Lantana*, though introduced not long ago as an ornamental garden plant, has spread throughout the island, and forms dense thickets wherever the land falls out of cultivation. The dispersing agents here are birds, which greedily eat the pulpy fruits. But we need not go so far afield as South America or Ceylon for our examples. A very striking instance of the ubiquity of germs, and the occupation by them of any suitable *nidus*, is seen in the almost invariable appearance of the Common Cord Moss (*Funaria hygrometrica*) on cinder paths, or places where the ground has been burnt. There may apparently be none of the Moss in the near neighbourhood : the ground has been sterilised by heat. It can only be that the spores are conveyed by currents of air, in which countless numbers are present, but relatively few find a spot for germination. The case is similar for the appearance of Moulds on bread recently baked, but kept in a confined atmosphere. Such

examples illustrate the magnitude, and the far reaching character of the results which may follow, in
cases where the immense potentialities conferred upon
organisms by the geometrical ratio of increase are
combined with effective means of dispersal. Most
plants possess some such potentialities, but few find
the opportunity of realising them.

The failure in so many cases leads us to enquire
what becomes of the enormous surplus of potential
lives which never get beyond the initial stage of the
germ? The fact is that the risks of youth are very
great indeed. Many germs are killed off almost at
once by unfavourable conditions, such as undue
temperature, or drought. Light affects some injuriously, and happily this is the case with many Bacteria.
Many fail to find a suitable *nidus* for germination ;
others succumb to unsuitable seasonal changes in the
defenceless condition of the seedling. Competition
crowds out others. Many fall victims to the predatory
attacks of animals, which naturally divert to their
own uses any food-stores designed for the germ.
Fungal attack accounts for the failure of others, and
especially in the seedling state. When we contem-
plate the multiplicity and the insistence of these risks,
the cause for astonishment is rather that any come
through the ordeal than that so many fail. There is,
however, a more occult series of obstacles to success
than those which have been mentioned, and though

they are less obvious they are no less potent in limiting a plant-population. I mean the inter-relations of organisms. Relatively few plants complete their whole life-cycle independently of other living things, and the degree of inter-dependence is often a surprise to beginners. Darwin in *The Origin of Species* cited the example of Red Clover and Cats. Having shown that the Clover is absolutely dependent upon Humble-bees for its pollination, without which no seed is set, he further pointed out that the number of Humble-bees depends in great measure upon the number of Field Mice in the district, which destroy their combs and nests. The number of Mice is dependent on the number of Cats, and he concludes that "it is quite credible that the presence of a feline animal in large numbers in a district might determine, through the intervention first of mice and then of bees, the frequency of certain flowers in that district." Again, it has been shown by elaborate observations in Europe and Asia that the northern limit of distribution of the Monkshood and that of the Humble-bees which pollinate their flowers, is almost exactly coincident, a fact which clearly points to an inter-dependence. Many similar instances might be quoted. But such relations are not confined to the higher plants : Goebel quotes the case of the Splachnaceae, those bright red and yellow Mosses which inhabit the dung and remains of animals. The question is how they attain with a high degree of certainty to this peculiar and circumscribed *nidus*, especially as their spores are aggregated in a sticky mass. It appears that Flies are the agents, and passing from station to station for the purpose of oviposition, they convey the sticky spores from old cultures of the Moss to the fresh dung. It is hard to see how this could possibly be effected without insect agency. Another case is that of the fungus, *Pilobolus*, which grows on the dung of horses and cows. When mature it can shoot off its sporangium with its sticky mass of spores to a considerable distance, and they adhere to grass or herbage in the vicinity. If this again be eaten by horses or cows, the spores passing through the alimentary tract will be ready to germinate upon the excreta, and so are very efficiently dispersed. These are examples of that inter-dependence which is so frequently a more or less necessary condition of the successful completion of the life-story. Such relations will go far to determine the numbers and spread of the species involved. It may thus happen that where the production of germs, their dissemination, and germination may have all been achieved, the species may still fail by reason of its inability, in some other respect, to complete the cycle of its life.

The general question which will present itself with some insistence is, what positive advantages following from the over-production of germs can repay the race for the vast expenditure of material upon them? A first result is the provision with a high degree of probability for the maintenance of the race in full numbers. The fisherman knows that if he takes a trout in a certain pool its place will be almost certainly occupied before long by another. Naturally such replacement of individuals cannot occur with any degree of regularity among the rooted plants of our area of sward. But the enormous over-production tends in this direction, and will suffice to secure a reasonable average over all. A second advantage gained is the probability of spread of the species to any new spot which is favourable, and the incursions of alien plants fully show how effective this may be, as soon as opportunity offers. A third result, and certainly not the least important, is that the overplus gives a basis for Natural Selection, bringing as a consequence the maintenance of the stamina of the race by the elimination of weaklings, and the preservation of favourable variations, with the consequent possibility of advance to a state of higher efficiency.

In late years there has been a definite tendency to minimise the effect of Natural Selection. It is hardly necessary to point out that Selection in itself is not a factor of advance. Naturally it can only contribute indirectly by the elimination of weaklings, and by sifting out favourable variations from among the vast number of potential lives. The ultimate question is the origin of those variations. Variants from the type are classified as (*a*) Mutations, and (*b*) Fluctuating Variations, and it is commonly held by those who are engaged with enquiry into such matters that the Mutations bring considerable divergence from the type and are transmitted to the offspring, while Fluctuating Variations involve as a rule minor divergence from the type and are not transmissible to the offspring. More especially has it been insisted upon that characters acquired during the individual life of the parent are not transmitted to the offspring. It has been shown, it is true, that certain considerable variations which are held as spontaneous Mutations are transmissible, and that certain minor variations rightly held as fluctuating variations are not transmitted. But what is not yet proved to the satisfaction of all is that no acquired characters are in whole or in part transmissible : nor is it clear that there is necessarily a distinction between minor mutations and fluctuating variations. The precise statement of the distinction between these two categories, and the assumption on the basis of a limited experience that the one is inherited by the offspring and the other not, has placed an obstacle in the way of understanding how direct adaptation to environment can possibly proceed. It is hard indeed to believe that all the wonderful accommodations of organisms to their surroundings are the product of blind chance. If there were any directive influence determining the course of inherited variation, and natural selection were operative upon the variations thus produced, the results which we see would be readily understood. At present such directive influence, traceable to the immediate accommodation of the organism to its environment, is directly ruled out by certain investigators. But there are many who hold their opinion in suspense on such questions. They look hopefully in the direction of a theory involving some form or other of the storing up of past experience, and its perpetuation in the features of the race. By selection among organisms with such memorised records the facts of adaptation to the environment would become intelligible. But of this the demonstration is not yet come.

Returning now, in conclusion, to the limited area of sward with its varied inhabitants from which we started this discussion, we shall see in it the result of a balance struck by Organic Nature. On the one hand is the fact of constant over-production of germs: on the other is the imminence of adverse fate threatening each life we see, while the field on which this tense battle of the individual is being fought is strewn with countless failures. Efficiency is the ruling condition of existence, and it is maintained by selection among multitudinous variants. If those variants are favourable, an ever-increasing efficiency is what such surroundings will naturally produce. The very plants we observe growing within our arms-length may even now be stamping upon their race new characters, which will form features of the vegetation of the future.

CHAPTER VIII

SAND DUNES

It has already been remarked that the sandy beach between tide-marks, being desert, is botanically uninteresting. But above high-tide level the seaward fringe of the Land Flora presents many features that may well arrest the attention. Where precipitous rocks face the sea there is little to remark beyond the straggling of the usual plants of the neighbourhood on to the slopes and ledges of the cliffs, but with a few plants added that are peculiar to the coast. Here may be found the Samphire, mentioned by Shakespeare as gathered on Dover cliffs, where it still grows and is collected for pickling. Or in Scotland the Lovage, which has been used as a pot-herb. The Wild Cabbage also, and the Carrot, Beet, and Sea-Kale occur here and there. In fact the Coastal Flora has given us the original stocks of many of our most ancient garden vegetables, the selection of them by our predecessors having doubtless been based upon the succulence of their foliage, which is in relation to their growth upon saline soil.

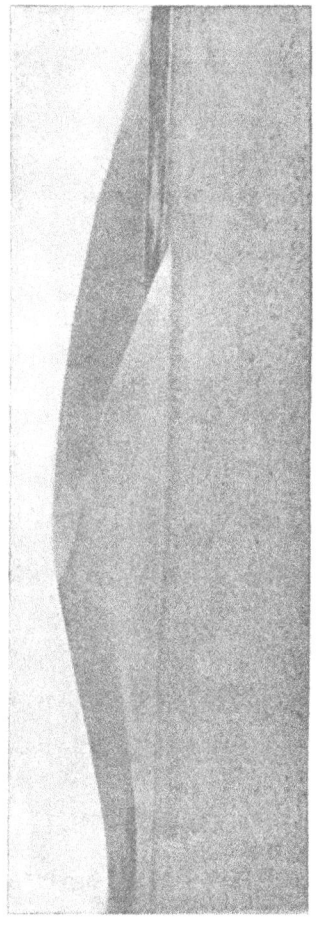

Fig. 19. A typical "Barchan" of the Libyan Desert: the Kharga Road Dune. (After H. J. L. Beadnell, F.G.S.)

A peculiar vegetation, however, appears in parts of the coast-line where a sandy beach leads up to a more level land-surface, and solid rock is absent, or at least discontinuous, or covered out of sight. There in the inlets and bays, or along considerable lengths of the shore, may extend those lands claimed by the Golfer as the only true "Links," protected seawards by rougher Sand-hills or "Dunes." Lastly, between these and the tide-marks there is usually a narrow, but more or less level expanse of dry sand, which may bear its own limited but interesting Flora. The substratum of the Links and Dunes is all composed of sand, with here and there it may be projecting bosses of the underlying skeleton of solid rock. The sand of the Dunes is entangled with rough Marram Grass. That of the Links is mostly covered by a continuous but thin sod, matted with a dense growth of smaller plants.

Sand-hills or Dunes may be formed by the action of the wind alone. Examples are seen in various localities, but they have been studied with special care in the Libyan Desert, where the wind-formed Dune takes a very definite crescentic shape styled a "Barchan" (Fig. 19). The moulding of the wind-formed Dune is commonly such that a gradual slope of 5—10 degrees on the windward side leads up to a ridge, from which the surface again falls with a slope of about 30 degrees on the leeward face. But

as this formation of ridges depends upon constancy
of direction of the wind, it is only occasionally
apparent on our own coasts, and the ridges themselves
are only of small dimensions; for here the variable

Fig. 20. Showing a stone, and a bunch of dead grass as seen on the
beach after a high wind. Each has a long streamer of sand
collected under its lee. In the close foreground are wind-ripples.

direction of the wind is liable soon to destroy such
constancy of form as a gale of some days may have
produced.

There has been some controversy as to the method

of the formation of these wind-created Dunes. The
prevalent view is that the sand hurried along by the
gale is heaped up over that already present till a
slight ridge appears, which is then constantly increased
by more and more being carried over it from the
windward side. The steeper angle of the leeward
face is then due to gravity acting on the grains which
lose pace on the sheltered side. This accounts for
the steeper slope, which advances constantly but
slowly down the prevailing wind. The plain fact is,
however, that,—putting such wind-formed Dunes on
one side since they do not form a leading feature
here at home,—the existence of definitely shaped
Sand-Dunes on our shores and the genesis and
maintenance of Sandy Links are essentially due to
vegetable life. It may accordingly be a matter of
interest to consider carefully how they originate.

The sand itself consists of the comminuted fragments of rock or of sea shells, and is cast up by the
waves upon the beach. Thence it is liable to be carried
by wind upward to the land. Once deposited in any
given position each grain, being unattached, is free
to be moved on again to another spot by any gust of
wind. But the probability of this happening in the
case of any individual grain is less if it should have
lodged in some place protected from the wind, while
it is more likely that the wind-borne particles will in
the first instance settle wherever there is a relatively

Fig. 21. Plants of Salt-Wort (*Salsola*), after exposure to strong wind blowing along the beach. Each plant has gathered around it a hillock of sand.

quiet spot. Such quiet spots are found under the
lee of any obstacle, such as a stone, a seaweed, or
a living plant. Any one who observes the beach after
a high wind, especially if it blew in a direction following the coast-line, will find innumerable instances
of the sand-streamers which thus collect under the
lee of any solid object, living or dead (Fig. 20). But
if the object itself be a living organism, and grows
upwards as the sand collects around it, then there is
the possibility of an actual building up of a permanent
heap, provided that the winds recur at intervals.
This is the principle which underlies the origin of
all permanent sand-formations on the coast, and the
conditions are sometimes repeated inland.   It is
the presence of certain sand-dwelling plants which
provides in the first instance the sheltered spots for
the settling of the sand, and secondly, it is by their
continued growth that that degree of permanence
and enlargement is secured which is characteristic of
the Dunes and Links as compared with mere shifting
sands.

We may take the formations thus built up from
the sand in the order in which they severally present
themselves as we pass upwards from the high-tide
mark.   First, there may be a zone of loose sand, dry
at the surface and easily shifted by the wind.
Spreading downwards into it, and not much, if at
all, raised above the beach, comes the advanced

guard in the form of plants belonging to various
affinities, but having this in common that they are
of low stature, and of more or less creeping habit.
They are characteristically "Halophytes," that is
denizens of saline soil, and most of them bear the
mark of it in their fleshy habit, their smooth hairless
surfaces, and their bluish tint due to a thin protective
covering of wax. Among such may be seen at the
lowest levels the Orache (*Atriplex*), and Seablite
(*Suaeda*), the Sea-Rocket (*Cakile*), the Salt-Wort
(*Salsola*) (Fig. 21), and Sea Purslane (*Arenaria pep-
loides*) : with these, or often forming dense patches by
themselves, are found the Sea-Couch Grass (*Triticum
junceum*), and the Common Sand Sedge (*Carex
arenaria*). These plants are joined not uncommonly
by some ordinary land-plants, such as the Silver Weed
(*Potentilla anserina*), and various forms of the Couch
Grass (*Triticum repens*). An interesting though rare
plant that shares this habitat is the pink-flowering
Convolvulus (*C. soldanella*) : it creeps along the sand
much in the same way as its relative, *Ipomaea pes-
caprae*, which is a notable sand-binder of tropical
shores. Such plants as these, or various representatives and combinations of them, by their mode of
growth give a coherence to the otherwise loose sand,
binding it together by their roots and runners. At the
same time, when wind is drifting the loose grains
along the shore, the shoots and foliage of the growing

B.   9

plants provide those local shelters in which the
moving sand collects.   But as many of these plants
are annuals, while none grow to any considerable
height, they are not efficient in securing permanence,
or in promoting any considerable rise of level of
sand collected.   The result is that the areas they
cover are usually flat, and slope directly down to the
general level of the beach.   Their effect is thus apt
to be local, and insignificant when compared with
that of the more efficient sand-binders of the Dunes.
Nevertheless the methods of action of these plants
are well worthy of observation and study, both from
the point of view of the self-preservation of the
plants under their peculiar surroundings, and as
regards the practical binding effect which they
produce.

Next in order landward come the Dunes, which
rise more boldly from the level of the beach, but as
often as not they abut directly upon the high water-
mark, and the lower growths just described may be
entirely absent (Fig. 22).   The Dunes that face the
sea are usually " White Dunes " : that is, the sand is
not covered in by vegetation, but is largely exposed,
so that the colour of the whole is that of the sand
itself rather than of the plants growing upon it.
They commonly present a steep face to the shore,
sloping more gently on the landward side.   This is
probably due to their growth seawards being re-

Fig. 22. Sand Dunes in Dunnet Bay, Caithness. The rounded coast-line is characteristic, with the grass-covered Dunes rising boldly from the level of the Beach. (After a photograph by H.M. Geological Survey for Scotland.)

peatedly checked by the high tides and heavy winter seas washing away their bases. The sand then falls in steep avalanches from above, and may be heaped up in various ways by the prevailing winds. But it is hardly possible to give more than a very general description of their form; it is so various. The factors that determine it are primarily the growth of the plants which serve to collect the sand and to bind it together ; and secondarily, the effect of the wind in again undoing what the plants have done, by carving out and redistributing the sand collected under their influence ; for there is in the White Dune formation a constant construction and as constant a destruction going on so long as the surface is not covered in by plant-growth. This absence of permanence suggested the designation of "Shifting Dunes."

Before we consider the plants which take a leading hand in the formation and the shaping of Dunes, it may be well to enquire what are the qualifications for their success. In the first place, Dune-forming plants must be perennials. Annual Dunes are always small and temporary, for when the plants die at the end of the season the sand collected round them is dispersed : this is the case with such plants as *Salsola* (Fig. 21). Secondly, the plants should be capable of spreading by means of runners, or by adventitious buds upon their roots. By this means the area of the Dune will be extended, and its growth to a greater height will be possible. A

Fig. 23. A dense mat of Marram Grass from which the sand has been washed away by the waves at high water, so as to show the horizontal rhizomes and the roots in their natural positions as they were embedded in the sand.

third point, which will give a peculiar advantage in Dune-formation, is the capacity of the plant for

growing upwards to the light when it may have been
completely covered by blown sand. The more rapidly
and effectively this can be accomplished the greater
will be growth of the Dune. Lastly, owing to the
low specific heat of the sand, and its free exposure,
and the consequent rapid cooling at night and liability
to sudden heat during the day, it will be necessary
that the plants of the Dune-flora shall be xerophyti-
cally adapted, that is, that they shall be able to stand
sudden conditions of extreme heat and drought. It
will be apparent that relatively few plants are in a
position to meet all these requirements, and it need
therefore be no matter for surprise that the Dune-
forming Flora is a restricted one.

By far the most common and effective Dune-
forming plant is the Marram Grass, or "Bent" as it
is called by golfers (*Ammophila arundinacea*). It
will therefore be worth while to examine the Bent
Grass carefully. It appears coarsely and densely
tufted above ground, with long leaves sheathing
below, but narrow in form, and very gracefully curved
in their upper region. They are furnished with
sharply pricking points, and doubtless these aid the
young growing leaves in making their way upward
through the sand. Examining the leaf it is seen to
have a convex surface (the lower or abaxial), which
is hard and smooth, and a concave surface (the upper
or adaxial), which is marked by deep longitudinal

furrows.   The behaviour of these leaves under varying circumstances of wind and moisture is interesting.
It may be noted on any breezy day that the leaves
present their smooth and convex surface to the wind,
so that not only is their mechanical resistance to it
the most effective, but also any sand it may convey
glances off instead of catching in the grooves of the
concave face.   Another point which is certainly an
adaptive one is the capacity of the leaf for rolling
in its margins in dry weather till it looks like a length
of green wire.   But when moisture is plentiful it
flattens out so that the smooth lower surface is almost
flat, and the grooved upper face is widely exposed.
This automatic change of form, which quickly alters
the general appearance of the grass after a shower,
is brought about by the contraction or expansion of
certain water-containing cells on the upper surface.
As these swell with moisture the leaf flattens, and
evaporation of water-vapour from the upper surface
is easy.   As the moisture becomes deficient the leaves
contract, the margins of the blade are drawn together,
and evaporation is checked.   Such automatic control
is one of those features which make life possible for
the plant under the exacting conditions of the sandy
Dune.   The inflorescence appears as a dense white
panicle in July, but it presents no characters of
special interest.   This much may be observed without
digging the plant up.   But below the surface of the

sand are features of still greater moment as bearing upon its value as a sand-binder.

Hidden in the sand are the long horizontally running rhizomes of the Marram Grass that are densely matted together, and are thus so marked a factor in its success (Fig. 23). On them the leaves are borne alternately at long intervals, and develope only their lower sheathing part while the lamina remains small. These rhizomes are found to be firm and resistant like strong cord, and their mechanical strength is again an important point. They may be traced for long distances, and are found throughout the structure of the Dunes, even to their base. At each node where a leaf is inserted a number of roots may arise, radiating outwards into the sand, while in an axillary position a bud originates, the mode of development of which may vary according to circumstances (Fig. 24). It may grow horizontally with long internodes, and become a new runner, having its terminal bud sharpened and indurated like the point of an awl : or it may turn upwards and form a leafy shoot, which on emerging at the surface of the sand will, by shortening its internodes and developing its axillary buds, form one of those dense leafy tufts so characteristic of the species. It is the adaptability of this grass to its circumstances in respect of direction of growth of its shoots, and the degree of elongation of its internodes, combined

with its xerophytic specialisation, that makes it before
all other plants the leading builder of Dunes.   Other
grasses take part also, for instance the Lyme-Grass

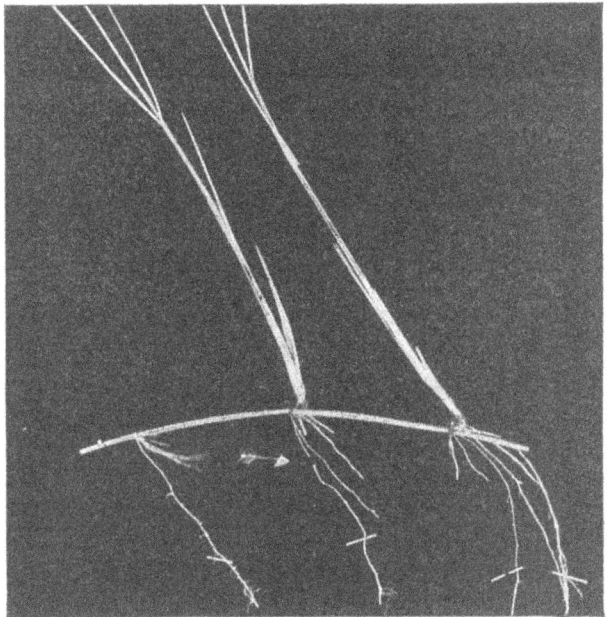

Fig. 24.   Horizontal rhizome of Marram Grass, showing three nodes,
from which roots arise.   From two of them leafy buds have grown
upwards through the sand.   The arrow points towards the apex
of the rhizome.

(*Elymus arenarius*), which shares many of the characters of the Marram, and the Sea-Couch Grass (*Triticum junceum*). But these are neither of them so successful, the latter together with the Sand-Sedge (*Carex arenaria*) being deficient in capacity for upward growth after being buried in the driven sand, in which respect the Marram especially excels.

The White Dune thus formed is sometimes styled the Shifting Dune. Having its sand retained only by entanglement amongst the relatively lax growths of the Marram, its surface is liable to be blown away again, while the sand removed from the windward side will be caught by the vegetation on the sheltered side. More especially does this occur where the Dunes have attained great height, for there not only are they more exposed to the wind, but also the plants, being higher above the level of the ground-water, are apt to grow less strongly, and form a less efficient protection. And so the whole Dune may appear to change its position, advancing by degrees down the prevailing wind. There is probably no better example of this in Great Britain than that of the sands of Culbin near Forres, on the south coast of the Moray Firth. Moreover the history of that strange waste of sand is so far known that an additional interest springs from it. The estate of Culbin, extending to some 3600 acres, was in the 17th century celebrated for its fertility. It was intersected by the river Findhorn,

and included, besides smaller holdings, some sixteen
farms. The disaster which buried the whole estate
beneath the sands came suddenly in the autumn of
1694. The sea had been encroaching extensively on
the land further to the westward, so that a large
expanse of sand was exposed at low water. It is
stated that a westerly gale carried immense quantities
of this sand over the fertile land, covering fields and
houses, burying the mansion-house itself, and even
diverting the course of the Findhorn river. Ever
since the Culbin Sands have been a vast shifting area.
If Marram Grass obtains a hold here or there, and
Dune-formation proceeds, it is again torn out by the
winds. A large part of the area remains as bare
Sand-hill, without a trace of vegetation upon it.
Here the surface sand is free to move with every
wind, and the form is constantly changing. The
effect of this is well shown at a point where a wood
of Scotch Firs had grown to maturity. The sand
advancing in a wreath some 40 feet high is gradually
enveloping the trees (Fig. 25). Elsewhere the sandy
pall is being removed again, and the dead stumps are
all that remains to show the destruction that has
been wrought (Fig. 26). The complete obliteration
of woods may be seen on even a larger scale elsewhere : for instance, on the shores of Lake Michigan.
Such effects illustrate to the full the irresistible
advance of sand borne as individual grains by a wind

constant in direction, but not necessarily great in power. The principle is the same as that which may be followed on any Dune of small size, where the wind has got the better of the binding vegetation.

Enough has been said to show the instability of the White or Shifting Dune. But provided that the conditions be quiescent, various plant-growths appear between the tufts of Marram or other prime formers of the Dune. These gradually cover the surface, converting the formation to what is called Grey or Stationary Dune. This constitutes a third zone on the landward side of the chain of White Dunes. The plants that bring this about are usually of low growth and straggling habit. Small annuals or biennials are found among them, such as Whitlow Grass, *Teesdalia*, and species of Mouse-ear : small Leguminous plants, such as *Lotus*, and species of Clover and Vetch. The Ladies Bedstraw, various small grasses, such as *Aira*, *Agrostis*, and Sheep's Fescue : or where in the hollows water will sometimes stand, species of Rush and Cotton Grass and Sedge may appear. Mosses may take their place in this attractive little Flora, some Fungi such as Puff-Balls and Clavarias, and Lichens such as *Cladonia*. But amongst the most effective plants which give a greater degree of consistence to the Grey Dunes are certain woody plants, such as the Creeping Willow (*Salix repens*), with its almost prostrate twigs, while the

Fig. 25. Landward margin of Maviston Sand-hills, showing a "shifting Dune" advancing and enveloping a well grown Fir wood. The tree-tops are seen projecting from the sand to the right. (After a photograph by H.M. Geological Survey for Scotland.)

little Burnet Rose (*Rosa spinosissima*) or the Rest Harrow (*Ononis spinosa*) may add a charm as well as a permanence to the formation. Of the larger woody growths which often follow the formation of permanent Dune, one of the most effective plants is the Sea Buckthorn (*Hippophae rhamnoides*), which is native in the south and is naturalised in Scotland. Not only is it xerophytic in its character, but it also fulfils the condition of spreading by the formation of adventitious buds on its long creeping roots. Lastly, such xerophytes as Gorse and the Heaths and Ling, make their appearance as well, and when once the Grey Dune is thus permanently established many other plants follow, and constitute a Flora of ever-increasing variety.

And so what is at first an unstable formation becomes compacted till it acquires a settled character. It must not, however, be thought that actual encroachment upon the sea itself is a common occurrence. Doubtless plant-growths, upon which Dune-formation depends, are continually edging seawards ; but they are liable to be as continually checked and curtailed by the action of the waves at high water in stormy weather. The methods of Dune-formation are not aggressive : they are rather those of defence, preventing inroads. Consequently Dunes are most commonly found to run in concave curves between headlands, and to round off the indentations of the coast-line,

Fig. 26. Remains of a Fir wood at Maviston Sand-hills, showing the dead trunks of trees previously buried, but now exposed again by removal of the "shifting Dune." (After a photograph by H.M. Geological Survey for Scotland.)

rather than themselves to project seawards. Like the softer tissues of the animal body which round off the more abrupt outlines of the skeleton, so the Dune-formation softens the harsher configuration of the rocky framework of the coast.

Nevertheless on certain low-lying coasts, which have no resistant skeleton of firmer rocks, and thus are liable to inroads of the sea, the defensive methods of the Dune-formers may be of the highest importance. Where a shallow, gently graded beach leads up to high-water mark, breaking thus the force of the waves even in stormy weather, the Dune-formation may oppose an effective barrier to further advance. Moreover, in such places the ample supply of sand that may readily be blown from the beach makes Dune-building easy. It is on certain stretches of the Continental Coast bordering on the North Sea that the Marram Grass and Lyme Grass attain their greatest use. They are planted out according to strict method, and are even protected by law. Thus Dune-formation may be not merely a picturesque incident of the Coast, but it may also be of positive use to man. In the next Chapter it will be shown that it may also minister to his amusements.

CHAPTER IX

THE game of golf finds its home at the sea-side. Widespread as it is to day inland, it was essentially of coastal origin. It was primarily a game to be played on a sandy substratum, and the constant effort of the inland green-keeper is to construct features which are characteristic rather of the sea-shore than of the midlands. However well he may carry this out, and even where a sandy soil favours his efforts, the inland course is a thing different from the true coastal links. Perhaps it is in part the tang of the sea, but assuredly it is in the main the contour of the surface and the quality of the turf, and of the sand below it, that make one realise that golf at the coast is the reality, and inland golf its mere shadow. Accordingly though this Chapter relates nominally to golf-links in general, it will apply particularly to coastal links as having those characters which stamped the game in its original form.

The links of the sea coast may be as various in conformation as the coast itself. In point of fact the rock-sculpture which underlies the surface gives them their primary characters. However thickly covered by sand and soil and varied vegetation, sooner or later the solid rock would be reached by boring, and it is in the inlets of the coast-line, and the depressions of this rock-surface, that the materials collect to form the links themselves. These may take the form either of towering Dunes, built up by agencies explained in the preceding chapter, or of more level sand-fields originating in various ways, but still owing their outlines primarily to the underlying rocks. Though often out of sight the rocky skeleton has thus been the prime factor in the shaping of the links, and the broader features of the course may for the most part be properly traced back to it.

Apart from the general conformation of the links thus defined by their rocky skeleton, the detailed characters, whether of contour or of surface, arise from the agencies which affect the transfer and lodgement of wind-borne sand. The mere existence of an inlet or hollow in the framework would tend to detain some of the sand blown upwards from the beach. But much more effective is the influence of perennial vegetation. Wherever this has obtained a footing sand-aggregates collect, and grow with its growth. The most potent of all such influences has doubtless been that of the Marram or Bent Grass, as explained in the foregoing Chapter. Let us hope that its work is properly appreciated by golfers. When a sliced drive lands a ball in the rough skirting the sea, instead of giving vent to objurgations on the Bents, it is well to remember that they are after all the best friends of the golfer, for without them the links, if they existed at all, would have been relatively flat and featureless. The rounded contours of the green, and especially those heights which have earned for themselves such picturesque local titles, and have caused so many moments of exultation or of black despair, owe their origin alike to this most effective of Dune-Formers.

But we have already seen that the White Dune which they at first produce is unstable, and liable to shift, partly owing to the isolation of the particles of the sand loosely held between the tufted leaves, partly to the failure of the Marram Grass itself as the Dune grows higher. It requires the growth of many smaller plants of various affinity, forming combinations as various as the coasts on which they dwell, to build up between the tufts of Marram Grass that dense felt which first gives permanence to the Dune. A sod is thus gradually compacted, which covers over the loose sand and protects it, converting the White or Shifting Dune into the Grey or Permanent Dune. Every Golfer knows how important is the continuity of this skin of turf, thin and delicate as it is. If even a small hole in it be left to itself it may be enlarged into an undesirable bunker by the action of the wind whisking away the loose sand that lies below. It is not without good reason that the replacement of a "divot" cut out by an iron club has become something more than a courtesy of the links. It should be accepted as an imperative rule, the observance of which is merely continuing the order of Nature. But the green-keeper, with his barrow of soil and packet of grass-seed, is doubtless a more efficient agent in the maintenance of the protective skin of fine turf: for the replaced sod frequently fails to settle again, especially where the turf is thin and sandy.

There are thus three factors which share the prime construction of the sea-side Links. The first is the rocky skeleton which defines the broad features of outline. Secondly, there are the Dune-Builders, by whose means the moving sand is temporarily held, and aggregated into heaps of various form: and thus are initiated those swelling contours which give so much of their character to the coastal courses. Thirdly, the skin of mixed vegetation which follows, giving permanence to the otherwise inconstant Dune. Jointly these factors supply that raw material from which the fully protected course

Fig. 27. Flood-lawns in rear of Sand-Dunes of Keiss Links, East coast of Caithness, with patches of *Salix repens*. (From a photograph by H.M. Geological Survey for Scotland.)

may be produced (Fig. 27). The further agents are
Animals and Man. Of Animals the most effective
to this end are rabbits and sheep. The former do
harm as well as good. It is true they nibble down
the coarser growths, keeping in check the exuberance
of certain of the woody plants that invade the links,
and do so much to give them permanence. They
crop the sward to an even velvet, stiff it is true
in the pile, but short and smooth such as the
Golfer loves, and seldom finds except on the true
Links of the coast. So close is this film of turf to
the underlying sand that as the ball falls upon it
from a long tee shot the sound is sometimes like
the ringing of a bell: veritably it is the sound of
the "Musical Sands." But while rabbits thus help
to produce a suitable turf, their holes and scrapings
are a perpetual trouble, and discount in appreciable
degree the advantages which they bring. Over and
above the irritation from loss of balls, and "cupped"
lies, there is the risk from opening up the underlying
sand to the action of the wind. The close nibbling
of sheep is also effective without these disabilities,
but they do not crop so close as rabbits. Both
agents frequently work together, but the rabbit is
there on sufferance rather than by choice, while
the sheep is penned upon the Links on purpose.

The Raw Material of the Golf Course, originating
as the resultant of the factors named above, and

trimmed by animal agency, presents many interesting
features for analysis. The varying characters seen
on divers shores may thus be traced back to their
origin through various conflicting influences. Here
the rocky substratum rises to the surface in projecting bosses, there it is more flat in form, and the
wind-borne sand covers it entirely out of view.
Close by an effete Dune with its grasses weakened in
growth is being attacked by the wind, and eroded into
hollows or "Blow-outs," which are however checked in
their spread by the covering of firmer skin compacted
of finer grasses, bedstraw, and thyme. Here may be
a strong stiff growth of whins, constantly edging
forwards and as constantly trimmed back at the
margin by persistent nibbling of animals. Other
spots may be roughened by heather, or the creeping
willow, or made scrubby by growths of elder or
other shrubs. Lastly, where an old "blow-out" has
occurred, eroding the sand almost down to the moist
water-table itself, the sour and muddy soil will
harbour a host of rushes and sedges, stiff and thick
in growth, and standing after rain ankle-deep in
"casual water."

Such is the material upon which the green-keeper
has to exercise his skill, whether constructive or
destructive. It is an ordered balance of Nature,
irregular enough in its results it is true, upon which
he enters as a disturbing influence. With his scythe,

his mowing machine, and his roller he does in mechanical and prosaic fashion what is so much more picturesquely done by the free agencies of Nature. He fixes the "blow-outs" as bunkers, often arresting the effect of the wind by lines of railway sleepers, or piles of sods, thus straightening their irregular edges into rigid lines. He opens out new and artificial hazards, placed not by accident as were those of the original course, but as intentional traps which bear their artificiality upon their very face. He cuts away the heather and creeping willow, trims down the bents, and reduces the green to the quality of the best-kept lawn. He may by these means produce a course on which a record score may be lowered by a stroke or two. But to the Naturalist he merely appears to upset the balance of Nature, the observation of which is one of the keenest delights of the links.

There have not been wanting those who have publicly expressed their regret at the change which is passing over some of our leading courses. The "Fairway" is mown and mown again. It widens imperceptibly as the process is repeated, till the outgoing course merges with the incoming. Not a Bent dares to show its head in these ever-broadening avenues. On some courses you must wander far afield indeed before any serious obstacle of vegetable origin penalises you. The policy of the green com-

mittee seems to be to subordinate the guidance of
the ball to mere long driving. But it is possible
that the future may see a reversal of that policy by
the reintroduction, or the special preservation, of
Bents, Heather, and Whins so as artificially to
restore some of those natural penalties that the
green-keeper has so ruthlessly removed, or have
disappeared, may be, owing to the ordinary traffic
of the Links. Though we may regret their loss,
we may still remember that however deeply the
artificialities of the modern upkeep of the Links
may affect the surface and its vegetation, they do
little to modify the main contours of the ground.
These with their irregular undulations, their arbitrary
slopes and towering heights, are the true product of
natural forces, and the determining factor that has
influenced their origin and their form more than any
other is the vegetation of long ago. The individual
plants that shaped the Dunes may have passed into
natural decay, or have been swept out of existence
by the modern green-keeper, but still their record
remains registered for ages in the undulating surfaces
of the Links.

CHAPTER X

GENERAL OUTLOOK ON THE FLORA
OF THE LAND

THE foregoing Essays are apparently disjointed, and aloof from one another. But just as a circle may be defined by three or more detached points, and be drawn so as to establish a clear relation between them, so the subjects of the preceding Chapters, isolated though they at first appear, may be drawn together so as to give some general outlook upon the vegetation around us. Ever since enquiry into the Origin of living forms took shape in theories of Evolution, the underlying principle has been recognised that it is from the simpler organisms that such origins are to be traced. The comparison of these with the more complex has been held to shed light on the genesis of the higher forms. On the other hand, certain conditions and methods of life are known to have led to simplification both of structure, and of the cycle of events in the completed life-story of the individual Thus parasitism leads to simplification of the organs of nutrition.

Further, among Fungi some forms appear to omit entirely certain stages which are shown in allied forms. The life-history may thus be curtailed and simplified. It is only with some limitation therefore that the thesis is to be accepted that the simpler organisms give a key to the problem of Descent. Subject to such necessary limitations, the comparative method may be adopted, and we may state a current theory of the Origin of the Vegetation of the Earth based upon such comparisons. As we do so we may recognise how readily it accords with the limited facts advanced in the foregoing pages.

A great body of evidence indicates that the origin of Plant-Life on Land was from aquatic surroundings. The water-problem is at the base of the whole physiology of Land plants. There is no requirement so imperative, and no adaptation so exact, as that for the supply and the control of the water which forms so large a constituent of any Plant of the Land. The stability of the plant while young, as well as the functional activity of each cell of its body during life, depend upon a balance of gain and loss of water being duly maintained. The relative ease of aquatic plants as regards their water-supply, compared with the almost desperate shifts of many terrestrial plants, suggests with some cogency that the latter is the derivative and specialised, the former the primitive habitat.

But far the most convincing evidence is that derived from the details of propagation. The simpler denizens of the water commonly show a phase where detached cells are motile in water, as we have seen in the case of *Ulva* and *Ulothrix* (Chapter II). Such an incident figures in the life-cycle of all the simpler Land plants. It is seen in Mosses, Liverworts, Ferns (Chapter III), Club-Mosses, and Horsetails. None of these live through their normal life without dependence at one critical point upon external fluid water as the medium for that motility. In this they are believed to show traces of their ultimate aquatic origin still preserved. It is not till we reach the Higher Flowering plants that freedom from that embarrassing condition of life is seen : for in them fertilisation is by a pollen-tube, and external fluid water is no longer necessary for the process. In this respect the Flowering plants have become in the fullest sense Plants of the Land.

Further than this it has been suggested (Chapter II) that the motile stage of Algae may itself have been actually the primitive state for them, and have preceded that non-motile phase which is usually styled the "plant." The ultimate ancestors of the Sea-weeds were probably detached and motile organisms, like the simple Flagellates of the present day. It is a very reasonable, and even physiologically a probable, view that primitive life was like them naked and unicellular, motile and unattached. The first step was probably that of encystment in a cell-wall. In some primitive forms this was soft and gelatinous, and motility was retained, as is seen in *Volvox*. But commonly the wall formed a firm coat, and non-motility naturally followed. This state proved effective since it conduced to protection, and offered no obstacle to self-nutrition, while it led readily enough to the constitution of organisms composed of more than a single cell. So long as such fixed and multicellular organisms grew submerged or between tide-marks, the further working out was not difficult. Since mechanical stability and conduction of water were not pressing necessities, such plants were free to attain great size, and this is fully realised in the larger Algae. Spread of the species was sufficiently provided for by water-carriage, either in the form of motile or non-motile germs. In fact a non-motile and attached habit of the whole plant presents few disabilities and definite advantages to plants growing in water. Hence the extent and the success of the Seaweed Flora as we know it.

But it was otherwise with the non-motile vegetation which, originating as we believe from some Algal source, spread to the Land, and developed into the terrestrial Flora as we see it to-day. Such plants are exposed in part or even wholly to an atmosphere seldom saturated with moisture, and for the most part relatively dry. A fixed habit is obligatory for them in order to secure their supply of water and soluble salts from the soil. The development of special tissues for conduction and for mechanical strength followed, and these are seen in a simpler type in the Mosses, but in quite an advanced condition in the Ferns and Horsetails. It was, however, on the propagative methods that the change of habitat from water to land brought special disabilities. The absence of external fluid water, except at intervals, put a check upon pro-pagation by self-motile germs. Fertilisation by spermatozoids motile in water could only take place at times of rain or copious dew. As a set off against this the formation of numerous detached spores, dry and dusty when ripe, and thus readily transferred by the wind, became a marked feature of this primitive vegetation of the Land. It is seen carried to high perfection in Mosses and Ferns (Chapter III). By this means, while sexual repro-duction in such plants was necessarily an uncommon event, a vast number of individuals might be produced as a consequence of a single fertilising act. This arrangement provided adequately for the maintenance of the race, and for the spread of its individuals. The provision for intercrossing was, however, deficient. It is true that there is an attraction of the motile sperm to the archegonium ; but the sphere of such influence is restricted by the limits of the water-medium within which it acts (Chapter III). Thus what may be styled the primary vegetation of the Land, including Mosses and Ferns and their allies, constituted a true Land Flora as regards their vegetative system, but their propagative method was conservative and ineffectual for land-living plants. It remained typical of their aquatic ancestry, and suffered from natural restrictions on dry land. Such a Flora might be styled in a sense amphibial.

It remained for the Seed-plants to break loose from these limitations, and to become a Land Flora in the full sense of the term. The most important change which they show is in the method of their fertilisation by the pollen-tube. The pollen-grains can be transferred from the stamen to the stigma at any time, and dry conditions will even favour the process, while the further growth of the pollen-tube down the style is independent of external fluid water. Thus the conservative restriction which hampered the primary vegetation of the Land was finally removed. It has been seen that the Mosses and Ferns and their allies are deficient in opportunity for intercrossing, as a consequence of their aquatic method of fertilisation. This disability also falls away in the Seed-plants. They show as one of their most popular and sensational features an extraordinary facility in the use of external agencies

for securing cross-pollination. The wind, occasionally
water, but chiefly animal agencies have been brought
into the service of intercrossing. The full armoury
of colour, of honey, and of scents, together with
elaborate mechanical devices, contribute to this end.
Their study has become almost a play-ground of
science. The flower has developed into a highly
specialised mechanism for the purpose of intercrossing.
First came the segregation of the nutritive and propagative regions of the originally non-specialised
shoot, by such steps as are illustrated among the
lower Vascular plants. The propagative region then
gradually assumed the features of the flower as it
is seen in such various forms at the present day
(Chapter IV). But while we recognise to the full
the beauty and the almost infinite variety of the
methods employed to secure cross-pollination, the
fact that stands behind them all, and is indeed their
ultimate justification, is the immobility of the plant,
so different from the ambulatory powers of the
higher animals. The plant cannot move as a whole
to seek its mate, and so if intercrossing is to take
place at all, external agencies must be pressed into
the service.

The same fact is the ultimate *raison d'être* of
the varied methods of dispersal of seeds and fruits
(Chapter VII). It has been seen how great is the
fecundity of many land growing plants. But the

full effect of this cannot be gained unless the spores and seeds be scattered, so that each shall have its individual chance on germination. The plant being immobile cannot effect this of its own motion. External agencies are made available by various adaptive features. It is again the wind, water-currents, and animal agents which are employed in overcoming this disability which follows so directly from the fixed position of plants. Thus we see a strong antithesis between the mobile animal and the immobile plant, which may be traced back ultimately to their differences of nutritive method. It is, however, between the higher terms of the two series that the distinction is most marked. Passing backwards to the simpler, and presumably earlier forms, they assimilate more nearly, till we find ourselves contemplating forms which occupy a borderland distinctive of neither kingdom, and suggestive of a common origin.

The spread of germs leads not only to the maintenance of the numbers of the species but also to the occupation of new sites. In fact vegetation, as we see it, may be held to represent a balance struck between the high capacity for increase of the species involved, and the restrictive influence of external circumstances. Examples have been quoted where the former obtains the upper hand, and a rapid spread of a species may be the result.

The contrary balance leads to restriction or even to extinction. It is in this light that the vegetation of the Golf-Links may be viewed. It may be taken as an example of the occupation of a barren area (Chapter VIII). The fresh blown sand from the beach is a *nidus* presenting from time to time a fresh surface, austere but ready to receive. Doubtless there is a natural selection among the many germs that fall upon it, and only those suited to the situation obtain a permanent hold. Putting the prime formers of the Dune on one side, the variety of the species that bring about the conversion of the Dunes from the White or shifting state to the permanent condition of the Grey Dune is great. The observation of it gives the opportunity for the study of a new Flora in the making, and it may be noted as we walk the Links, how many are the species whose germs are called into existence, and how few are physiologically chosen.

In drawing this Series of Sketches to a close, it must be clearly understood that no attempt has been made to follow with exactitude the Evolutionary History of Plants as we see them. What has been attempted has been to illustrate along various lines of thought, each suggested by common features of the country, the outlook of Modern Botany. Each of these lines converges in one way or another upon the central problem of descent. The most prominent

result is to bring into a clear light the importance of
the relation of plants to water. In particular, attention
has been drawn to the water-problem as it affects
Land-plants. It is indicated that the acute incidence
of it is a consequence of the migration of organisms
originally aquatic to the land.   In fact, a broad comparison of vegetation as a whole shows that the
Flora of the Land is not primary, but secondary.
Its constituents have found it necessary to adapt
themselves to the atmospheric surroundings which
they have adopted. This has not always been successfully carried out. The adaptation has sometimes
been hampered by an almost inexplicable conservatism.
It is the study of such features which has given the
clearest clues. In them we may find evidence, still
preserved, of that greatest of migrations in an age
long past, viz. the transition of Plant-life from Water
to the Land.

INDEX AND GLOSSARY

For EU product safety concerns, contact us at Calle de José Abascal, 56–1°,
28003 Madrid, Spain or eugpsr@cambridge.org.

www.ingramcontent.com/pod-product-compliance
Ingram Content Group UK Ltd.
Pitfield, Milton Keynes, MK11 3LW, UK
UKHW010850090126
466816UK00011B/143